HAZARDOUS ORGANIC POLLUTANTS IN COLORED WASTEWATERS

HAZARDOUS ORGANIC POLLUTANTS IN COLORED WASTEWATERS

NATALIJA KOPRIVANAC
AND
HRVOJE KUSIC

Nova Science Publishers, Inc.
New York

Copyright © 2009 by Nova Science Publishers, Inc.

All rights reserved. No part of this book may be reproduced, stored in a retrieval system or transmitted in any form or by any means: electronic, electrostatic, magnetic, tape, mechanical photocopying, recording or otherwise without the written permission of the Publisher.

For permission to use material from this book please contact us:
Telephone 631-231-7269; Fax 631-231-8175
Web Site: http://www.novapublishers.com

NOTICE TO THE READER
The Publisher has taken reasonable care in the preparation of this book, but makes no expressed or implied warranty of any kind and assumes no responsibility for any errors or omissions. No liability is assumed for incidental or consequential damages in connection with or arising out of information contained in this book. The Publisher shall not be liable for any special, consequential, or exemplary damages resulting, in whole or in part, from the readers' use of, or reliance upon, this material.

Independent verification should be sought for any data, advice or recommendations contained in this book. In addition, no responsibility is assumed by the publisher for any injury and/or damage to persons or property arising from any methods, products, instructions, ideas or otherwise contained in this publication.

This publication is designed to provide accurate and authoritative information with regard to the subject matter covered herein. It is sold with the clear understanding that the Publisher is not engaged in rendering legal or any other professional services. If legal or any other expert assistance is required, the services of a competent person should be sought. FROM A DECLARATION OF PARTICIPANTS JOINTLY ADOPTED BY A COMMITTEE OF THE AMERICAN BAR ASSOCIATION AND A COMMITTEE OF PUBLISHERS.

LIBRARY OF CONGRESS CATALOGING-IN-PUBLICATION DATA
Available Upon Request

ISBN: 978-1-60456-936-0

Published by Nova Science Publishers, Inc. New York

CONTENTS

Preface		vii
Chapter 1	Introduction	1
Chapter 2	Colored Wastewater Treatment by AOPs: A Review of Recent Studies	9
Chapter 3	Study of C.I. Reactive Blue 137 Wastewater; Treatment by Several AOPs	31
Conclusion		55
Acknowledgement		57
Reference		59
Index		73

PREFACE

The manufacturing and the application of organic dyes involve the production and the handling of many organic compounds hazardous to human health. Many of these substances are considered toxic, even carcinogenic. Over the past couple of decades, manufacturers and users of dyes have faced increasingly stringent legal regulations promulgated to safeguard human health and the environment. So, there is a clear need to treat dye wastewater prior to discharge into the primary effluent. The limitation of traditional wastewater treatment technologies (biological and physical), such as low rate, disability to degrade many of recalcitrant organic dyes and the production of secondary waste which demands further treatment, can be overcome by the utilization of advanced oxidation processes (AOPs). These wastewater treatment methods are considered as low- or even non-waste generation technologies. AOPs are based on the production of very reactive species, such as hydroxyl radicals, able to decolorize and to reduce recalcitrant colored wastewater loads due to the high oxidation power and the lack of selectivity of OH radicals towards a broad range of organic pollutants present in wastewater. Generally, AOPs can be broadly classified concerning the way of OH radicals generation into chemical, photochemical, photocatalytic, mechanical and electrical technologies. This research deals with the application of chemical and photochemical AOPs for the minimization of recalcitrant colored pollutants present in wastewaters. The comparative study of several processes, Fe^0/H_2O_2, $UV/Fe^0/H_2O_2$, UV/O_3 and $UV/O_3/H_2O_2$, was performed through an experimental research of decolorization and mineralization of a reactive azo dye C.I. Reactive Blue 137, as a model wastewater pollutant. Applied processes were optimized according to their process parameters, Fe^0 dosages, initial H_2O_2 dosages and initial pH values of treated solutions. The influence of initial organic dye concentration, as well as the addition of solid particles, synthetic zeolites, on the process effectiveness was also investigated. Studied AOPs were evaluated on the basis of their eco-effectiveness, by the means of color (A_{610}), aromaticity (UV_{280}), TOC and AOX value decrease, and their cost-effectiveness as well.

Chapter 1

INTRODUCTION

COLORED WASTEWATER

Dyes make our world beautiful, but they bring pollution. Synthetic dyes are present in many spheres of our everyday life and their application is continuously growing. Organic synthetic dyes have been widely used as colorants in different industries such as textile, paper, color photography, pharmaceutical, food, cosmetic, electric... [1]. More than 0.7 million tons of organic synthetic dyes are produced annually worldwide. In additon, over 10,000 different dyes and pigments have been applied in those industries. Studies indicated that approximately 15 % of produced synthetic dyes per year have been lost during manufacturing and processing operations that involve the production and handling with many organic compounds hazardous to human health. Wastewaters originated from dye production and application industries present a very serious environmental problem because of the aesthetic nature due to the fact that the coloration is visible even in a low dye concentration. Although the presence of color in water might not appear to have a negative effect other than aesthetic, this is not actually the case. The color can adsorb and reflect sunlight entering the water. Consequently, bacteria can not grow sufficiently in the water and that can have a disastrous effect on the food chain [2]. But even more important, many substances in such wastewaters are considered toxic, and even carcinogenic and mutagenic [3]. Nowadays, the mostly used group among all dyes are reactive synthetic dyes, mostly applied for dyeing of natural fibers such as cotton and wool. Due to the high solubility and hydrolysis as a side effect of such types of dyes, they can often cause loaded colored effluents [4]. Almost 70 % of all reactive dyes are of the azo type, while other are mainly of anthraquinonic and phtalocyanine types [5]. Furthermore, commercial reactive azo dyes are intentionally designed to resist biodegradation, i.e. the wastewater originated from reactive dye processes is characterized by poor biodegradability, passing unaffected through conventional treatment systems and discharging into the environment [6-8]. Like other colored wastestreams, those from reactive azo dye manufacturing and utilization industries pose a major threat to the surrounding ecosystems due to the documented health hazards caused by the toxicity and the potentially carcinogenic nature of such organic pollutants [9]. The group of most dangerous contaminants in such wastewaters are aromatic amines which may cause methemoglobinemia, and they are generated as by-products of partial degradation of reactive dye wastewater treatment or as impurities present in commercial azo dyes due to their usage in production process. The amines oxidize the heme iron of hemoglobin from Fe(II) to Fe(III), blocking the oxygen binding. This results in characteristic symptoms like cyanosis of lips and nose, weakness and dizziness. [10]. A number of azo dyes have been shown to be animal carcinogenous, and analysis based on structure-activity relationships suggests that many more azo dyes are likely to have a carcinogenic potential. Reductive cleavage to carcinogenic aromatic amines, which are known to act via genotoxic mechanisms, is in all likelihood an important mechanism of azo dye carcinogenicity [7]. Since 20 years ago some intermediates for azo reactive dye manufacturing, such as: *o*-aminoazotoluene, *p*-aminoazobenzene, *o*-anisidine, 2-amino-4-nitrotoluene, benzidine, 2-naphthylamine, *o*-toluidine, etc, were become to EPA's "black list" of hazardous chemicals that resulted with their restriction. But today, many amines are still in use, resulting with their presence in wastewater originated from azo reactive dye manufacturing and application industries.

WASTEWATER TREATMENT METHODS

Due to the above mentioned facts, manufacturers and users of reactive dyes have faced, over the past two decades, increasingly stringent legal regulations promulgated to safeguard human health and the environment [5]. To comply those requirements, the new technologies have been designed, specifically, to analyze and remove color and priority pollutants from wastewater effluents and the methodology of the cleaner production has been applied to circumvent pollution problems by eliminating their source [11].

Generally, wastewater treatment methods can be broadly classified into biological, physical (i.e. mechanical) and chemical [3, 12-14]. Biological treatment methods are widely applied for the treatment of municipal and industrial wastewaters. Although biological treatment methods have many advantages (economical acceptance, simplicity of treatment facility), they can not be successfully applied for the treatment of colored wastewater containing reactive dyes due to their poor biodegradability caused by recalcitrant nature [8, 14]. Reactive dyes are developed and designed to resist fading upon exposure to light, water and oxidizing agent [15, 16]. They are characterized by low biomass adsorbability and they are not degradable by aerobic conditions in conventional biological wastewater treatment facilities. Recently, some microorganism cultures, which are able to decolorize and mineralize some of reactive azo dyes, were developed [17-19]. However, biological processes by itself are often inappropriate due to the low rates of biological degradation. Physical, also known as mechanical, treatment methods such as: filtration, sedimentation, adsorption, coagulation/flocculation etc., like their name states, include only physical/mechanical processes, without chemical transformations of present contaminants, for wastewater treatment. Therefore, they generally transfer waste components from the liquid phase to a sludge phase. That makes them insufficiently effective and moreover, very expensive, due to the requested further treatment of generated secondary waste or due to the regeneration of inactive adsorbents. Although there are many examples in the literature about textile wastewater treatment by activated carbon adsorption with reported process effectiveness for wide range of textile dye types, the relatively low adsorption capacity of activated carbon for some of reactive dyes was shown [20]. Reactive dyes can be also removed from wastewaters by adsorption on various natural quaternized organic materials such as cellulose, sugarcane bagasse, rice husk, and coconut husk, alternatively cheap adsorbents. But here, too, no successful regeneration has been reported [21]. Reactive dye removal by adsorption onto mezoporous minerals such as sepiolites and zeolites was also studied, but successful removal was limited due to the low adsorption capacities of these natural minerals [20]. Removal by synthetic zeolites is possible but it depends on pore size and capacity level of applied zeolites [22-24]. Coagulation/flocculation process using Fe(III) and Al(III) salts, or polymeric flocculant Levafloc-R showed high efficiency for decolorization of wastewater containing reactive dyes [25-27]. The main problems of this relatively cheap and easily to maintain treatment method are related with the generation of large amounts of sludge as secondary waste that requires further treatment and disposal. Chemical treatment is based on the electron transfer, and most widespread application found processes using chlorine, ozone and potassium permanganate. The main disadvantage of chlorine usage for wastewater treatment is the production of chlorinated hydrocarbons that are in most cases even more toxic and carcinogenic than parent contaminants and therefore the usage of chlorine is restricted in many countries [28, 29]. Besides that, the usage of chlorine as an oxidant for wastewater treatment is characterized by its low selectivity and high consumption. The application of pure ozone oxidation is limited by high selectivity and slow kinetics. Although ozone was shown as very efficient for decolorization of colored wastewaters, it rarely produces complete mineralization to CO_2 and H_2O [30]. The usage of potassium permanganate as rather expensive oxidative agent is widespread due to the easiness of its application. But the main problem is caused due to the production and the precipitation of manganese oxide which than should be removed from the process equipment. From the previous statements concerning wastewater treatment methods it could be concluded that common treatment methods suffer from various limitations and that higher efficient treatment methods need to be found.

ADVANCED OXIDATION PROCESSES

Advances in chemical treatment of wastewater resulted with the development of a number of alternative chemical technologies; so-called Advanced Oxidation Processes (AOPs). In last two decades, the interest of not only world scientific community for the development and the application of AOPs as wastewater treatment methods was rapidly increased. The rapid increase of interest of research on AOPs resulted with numerous scientific publications, but also with their numerous applications on full-scale treatment facilities [31]. The main advantage of AOPs over other wastewater treatment processes; biological, physical-mechanical and classical chemical is their pronouncedly destructive nature that results with the mineralization of organic contaminants present in wastewater. AOPs are considered as low- or even none-waste generation technologies [32, 33], and generate short-lived chemical species with high oxidation power. At the first place, that are hydroxyl radicals, (HO•), which Glaze et al. [34] pointed out as main oxidation species in AOPs. Owing to high reactivity (table 1) and unselectivity (table 2) of HO•, AOPs are considered to be promising methods for the treatment of hazardous toxic organic pollutants in aqueous solutions [35]. AOPs generate OH• in sufficient quantities to oxidize the majority of organics present in the effluent water [36]. However, some of the simplest organic compound such as acetic, maleic and oxalic acid, as well as acetone, chloroform and tetrachloroethane can not be oxidized by OH radicals [37].

Table 1. Standard redox potential of some oxidant species [35]

Oxidant	Redox potential, $E°$, V
Flour	3.03
Hydroxyl radical	2.80
Atomic oxygen	2.42
Ozone	2.07
Hydrogen peroxide	1.77
Permanganate ion	1.67
Chlorine	1.36
Chlorine dioxide	1.27

Table 2. Chemical species those are able to be oxidized by hydroxyl radicals [36]

Acids	formic, gluconic, lactic, malic, propionic, tartaric
Alcohols	benzyl, *tert*-butyl, ethanol, ethylene glycol, glycerol, isopropanol, methanol, propenediol
Aldehydes	acetaldehyde, benzaldehyde, formaldehyde, glyoxal, isobutyraldehyde, trichloroaldehyde
Aromates	benzene, chlorobenzene, chlorophenol, PCBs, phenol, hydroquinone, catechol, benzoquinone, *p*-nitrophenol, toluene, xylene, trinitrotoluene
Amines	aniline, cyclic amines, diethylamine, dimethylforamine, EDTA, propanediamine, *n*-propylamine
Dyes	azo, anthraqiunone, triphenylmethane
Ethers	tetrahydrofuran
Ketones	dihydroxyacetone, methylethylketone

AOPs show high flexibility in their practical application due to the fact that it can be used either separately or in combination, or in combination with other classical wastewater treatment methods. Furthermore, another great advantage of AOPs over classical wastewater treatment methods is the possibility of conduction at ambient conditions, i.e. atmospheric pressure and room temperature. Such so-called ambient AOPs can be used for the treatment of low loaded wastewaters (Figure 1) [38]. For the treatment of highly loaded wastewaters, even up to value of COD (chemical oxygen demand) ecological parameter of 200 gL^{-1}, other processes generating also OH radicals can be used. Some authors consider those processes as AOPs, although these processes require pressure and temperature highly over ambient conditions, so they can be named as non-ambient AOPs. Such processes are wet air oxidation, supercritical wet air oxidation and other hydrothermal oxidation processes that requires pressures and temperatures even above 1 MPa and 150 °C, respectively [33, 35, 38].

Regarding to the value of TOC (total organic carbon) content, AOPs are suitable for the treatment of wastewaters loaded in range of 100-1000 mgC L^{-1} [39]. Ambient AOPs can be broadly classified concerning the way of OH radical generation. Generally, OH radical can be generated by chemical, electrical, mechanical or radiation energy. Therefore, AOPs can be classified into chemical and catalytic, photochemical and photocatalytic, mechanical and electrical processes (Figure 2.).

Chemical processes involve the application of ozone and/or hydrogen peroxide, while a subcategory of this type of AOPs can be named catalytic processes that involve usage of some powerful catalyst (e.g. iron or cupper ions) in combination with hydrogen peroxide to produce OH radicals, so-called Fenton type processes. Photochemical and photocatalytic processes involve application of UV or solar irradiation in combination with some powerful either oxidant (ozone and/or hydrogen peroxide) or photocatalyst (e.g. TiO_2, ZnO, etc.). OH radical can be also produce under influence of mechanical (e.g. ultrasound process, radiolysis) or electrical (e.g. electrohydraulic discharge and non-thermal plasma processes) energy [31-40].

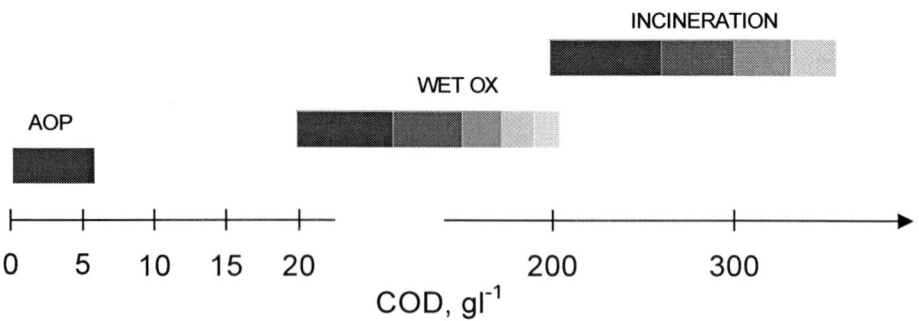

Figure 1. Suitability of water treatment technologies according to COD contents [38].

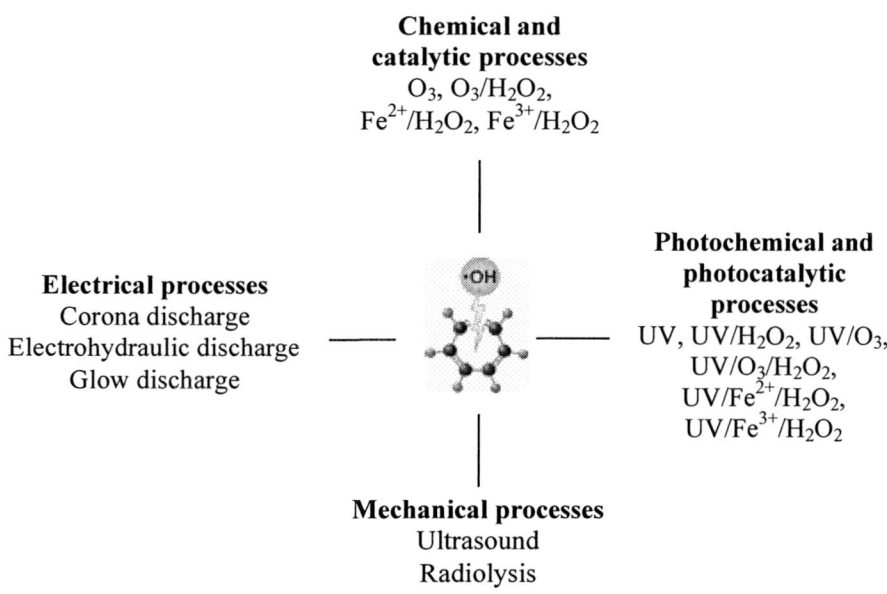

Figure 2. Schematic of advanced oxidation processes classification.

Chapter 2

COLORED WASTEWATER TREATMENT BY AOPS: A REVIEW OF RECENT STUDIES

According to the number of AOPs investigations directed to the treatment of colored wastewater, the most studied processes are Fenton type processes, then UV-based technologies, ozone-based technologies, etc. This part of chapter contains the more detailed review of AOPs applied for the treatment of colored wastewater investigated in our laboratory, and the less detailed overview of other AOPs.

FENTON TYPE PROCESSES

All processes that involve catalytic reaction between hydrogen peroxide and metal homogeneous ions, in the most cases Fe^{2+} ions, can be named as Fenton type processes. The reactivity of this system was first observed by Fenton in 1876 [41, 42]. Its utility was not recognized immediately, but 1930s when its mechanism based on OH radicals was proposed, the number of investigations concerning the application of Fenton reagent for the degradation of organic compounds dissolved in water is continuously growing. The application of Fenton process is based on the reaction that involves the oxidation of Fe^{2+} ions to Fe^{3+} ions with H_2O_2 and the subsequent generation of OH radicals (1):

$$Fe^{2+} + H_2O_2 \rightarrow Fe^{3+} + HO\bullet + HO^- \qquad (1)$$

Such generated OH radicals could then react with organic matter present in water. Although reaction shown by equation (1) is often referred as Fenton reaction [43], and presents the key step in the Fenton process, other important reactions also occur. The following set of reactions describes the occurrence of Fenton catalytic cycle:

$$Fe^{2+} + HO\bullet \rightarrow Fe^{3+} + HO^- \qquad (2)$$

$$Fe^{3+} + H_2O_2 \rightarrow Fe^{2+} + HO_2\bullet + H^+ \qquad (3)$$

$$Fe^{2+} + HO_2\bullet \rightarrow Fe^{3+} + HO_2^- \qquad (4)$$

$$Fe^{3+} + HO_2\bullet \rightarrow Fe^{2+} + O_2 + H^+ \qquad (5)$$

if the organic matter, i.e. organic pollutant, is present in the system, the next reactions are also involved in chain radical mechanism:

$$RH + HO\bullet \rightarrow H_2O + R\bullet \qquad (6)$$

$$R\bullet + Fe^{3+} \rightarrow Fe^{2+} + products \qquad (7)$$

By-products of above reaction (7) could be further degraded by radical mechanism to the complete mineralization. Besides Fenton process, Fe^{2+}/H_2O_2, that involves the application of ferrous salts, mostly ferrous sulphates, as a source of iron catalyst for Fenton reaction (1) [44-49], there is a number of studies that investigated the application of so-called Fenton "like" processes for the degradation of pollutants in wastewater. There are three types of Fenton "like" processes. The first group of processes considers the usage of ferric salts instead of ferrous salts as catalysts for the incitation of Fenton reaction (1) [48-51]. Next group of processes considers the usage of heterogeneous Fenton type catalysts such as iron powder, iron-oxides, iron-ligands, or iron ions doped in zeolites, pillared clays or resins, instead homogenous ferrous ions obtained from the dissolution of added ferrous salts [48, 49, 52-62]. Third group are processes that use other metal ions, e.g. cupper, manganese or cobalt, instead of ferrous ions in Fenton reaction [44, 63]. The primary benefits of Fenton type processes are their ability to convert a broad range of pollutants to harmless or biodegradable products and the fact that their relatively cheap reagents are safe to handle and are environmentally benign [43]. Even though these systems offer a cost effective source of HO•, their efficiency is limited with a couple of disadvantages: (a) the need for the removal of remaining iron ions after the treatment and (b) a limited yield of reaction process due to the formation of stable Fe^{3+}-complexes [37, 43]. Those limitations can be overcome by the usage of heterogeneous Fenton-type catalysts which can lower the final concentration of iron ions in the bulk after the treatment, while by the assistance of UV irradiation the formed Fe^{3+}-complexes can be destroyed, thus allowing Fe^{3+} ions to participate in the Fenton catalytic cycle [60, 64]. The oxidation power of Fenton type processes depends of several operating parameters such as iron concentration, source of iron catalysts, or catalyst if other metal ions are used instead of iron ions, H_2O_2 concentration, applied catalysts/hydrogen peroxide ratio (Fenton reagent), temperature, pH and treatment time [65]. According to the literature [65], a minimal threshold concentration of ferrous ions that allows the reaction to proceed within the reasonable period of time ranges between 3-15 mgL^{-1}, regardless of the concentration of organic pollutant. However, iron levels <25-50 mgL^{-1} can require excessive reaction times (10-24 hours). Typical Fenton catalyst/H_2O_2 ratios range between 1 : 5 to 1 : 25 [65, 66], while it is markedly proven that the optimal pH range for Fenton process is between 2.8 and 3.0 [43-54].

As it was mentioned above, Fenton type processes are the most studied processes for the treatment of colored wastewater. Numerous studies dealt with the application of Fenton process Fe^{2+}/H_2O_2 for the decolorization of colored wastewater including those that contain reactive azo dyes. Xu et al. [44] investigated the efficiency of Fe^{2+}/H_2O_2 for the decolorization and the mineralization of 20 organic dyes including 3 reactive dyes and they obtained rather high decolorization, >96 %, while mineralization extents after 120 min of treatment ranged between 72-79 % TOC removal. Lucas and Peres [45] achieved bleaching between 83 and 96 % depending on the initial dye concentration. Meric et al. [46] applied Fe^{2+}/H_2O_2 process for the decolorization and the detoxification of three reactive dyes with the initial dye concentrations of 100 and 200 mgL^{-1}. The overall decolorization obtained in that study was >98 %, while COD removal yielded between 92 and 95 %. In our previous study [49], Fenton process was applied for the degradation of two reactive dyes (azo and anthraquinonic type), and the bleaching >95 % was achieved in cases of both reactive dyes, while final extents of partial mineralization, 34.3 and 72.1 % of TOC removal, depended on dye structure. Neamtu et al. [51] obtained >99 % of color removal for both studied reactive dyes with the initial concentrations of 100 mgL^{-1}. Hsueh et al. [67] investigated the efficiency of Fe^{2+}/H_2O_2 using the low concentration of ferrous salts for the decolorization of reactive dye model solution, and they obtained similar results as in previous cases, rather high decolorization, >96 %, and <20 % of TOC removal after a one-hour treatment. From the above presented results of several studies applying Fe^{2+}/H_2O_2 process for the degradation of reactive dyes in wastewater, it can be observed that rather high decolorization was achieved in all cases, while according to the other monitored ecological parameters, i.e. TOC or COD, in all studies the mineralization of colored wastewater was only partial. Similar results were obtained by using ferric salts instead of ferrous salts [49, 67]. Several studies investigated application of heterogeneous catalysts for Fenton reaction and obtained similar results. Decolorization >98 % and mineralization extents of 45.3 and 46.9 % TOC removals for C.I. Reactive Blue 137 and C.I. Reactive Blue 49 model wastewaters, respectively, were obtained in our previous study by using iron powder instead of homogenous ferrous ions [49]. Tang and Chen [54] applied iron powder/H_2O_2 system for the treatment of colored wastewater containing C.I. Reactive Red 120, and obtained color removal >90 %. They pointed out iron powder/H_2O_2 system as effective for dye removal owing to the Fenton reaction that occurs in the bulk and on the iron powder surface. Several studies include the application of ferrous or ferric ions doped in zeolites or resins, as well as the usage of ferrioxalate, for the decolorization of model wastewaters containing reactive dyes [51, 60, 62, 69, 70]. By these heterogeneous Fenton type processes rather high dye degradation was estimated, and mostly owing to the occurrence of Fenton reaction on the surface of heterogeneous catalysts resulting with lower concentration of iron ions in the bulk after the treatment. But, in most of those studies the reactivity of Fenton type processes were improved by UV irradiation or by non ambient conditions (so-called wet oxidation process). Presence of UV light in Fenton type processes could give them several benefits. One of benefits is an additional source of OH radicals besides primary source throughout Fenton mechanism. Therefore, OH radicals could be generated in photo-Fenton processes from the photolysis of hydrogen peroxide, equation (8), and from the reduction of ferric ions to ferrous ions, equation (9). Furthermore, UV light could provide the avoidance of breaking the Fenton catalytic cycle due to the formation of stable Fe^{3+}-complexes between free Fe^{3+} ions and some aliphatic acids formed as by-products of dye degradation, shown by equation (10) [44, 45, 51, 62, 64, 68].

$$H_2O_2 \xrightarrow{h\upsilon} 2HO\bullet \qquad (8)$$

$$FeOH^{2+} \xrightarrow{h\upsilon} Fe^{2+} + HO\bullet \qquad (9)$$

$$Fe^{3+}(L^-) \xrightarrow{h\upsilon} Fe^{2+} + L\bullet \qquad (10)$$

Also, the presence of UV irradiation could enable these processes to achieve the complete mineralization due to the degradation of some OH radical persistent by-products, such as oxalic and acetic acid, by UV light [64, 69-71]. Throughout all these studies it was found that the decolorization of reactive dye wastewater by Fenton type processes was rather high and fast. On the other hand, only partial degradation/mineralization of formed organic by-products after initial cleavage of dye chromophores indicate high susceptibility of dye chromophores to OH radicals in the comparison to the remained organic part of colored wastewater. However, Fenton type processes could be proposed as one of successful treatment methods for colored wastewaters containing reactive dyes.

UV-BASED PROCESSES

Under the term "UV-based processes" are considered all processes that apply UV light either for the degradation of organic pollutants or for the initiation of oxidation mechanisms by the irradiation of some powerful oxidants or photo-catalysts, throughout then will be provided degradation of organic pollutants present in water. Hence, UV-based processes could be classified into photochemical processes, where UV light is used for the photolysis of some powerful oxidants such as H_2O_2 and/or O_3, and photocatalytic, where UV light is used for the irradiation of some powerful photocatalyst such as TiO_2, ZnO etc.

UV Photolysis

Since 1666 when Isaac Newton observed the diffraction of white light beam when passing through a prism, the investigations regarding UV light and its properties were began. At the begging of 19th century, the radiant energy beyond two ends of the spectra of visible lights was discovered. One of those is identified as infrared and another as ultraviolet region. Furthermore, it was shown that invisible chemically active irradiation beyond violet end of spectrum were the subject of laws of interference. All those observations, along with further investigations, indicated to the fact that irradiation with visible (VIS), infrared

(IR) or ultraviolet (UV) light has characteristics of the same electromagnetic irradiation, but they differ regarding to its frequency, and, what was discovered latter, pertaining energy [35, 72].

Table 3. Radiation type and pertaining energy [35]

Radiation	Wavelength, nm	Energy range, kJ Einstein^{-1}
IC	>780	<155
VIS	780 – 400	155 – 300
UV-A	400 – 315	300 – 377
UV-B	315 – 280	377 – 425
UV-C	280 – 100	425 – 1198

Therefore, UV radiation comprises energies from about 300 kJ Einstein^{-1} (UV-A radiation, 1 Einstein = 1 mol of photons), up to 1200 kJ Einstein^{-1} (vacuum UV). Table 3 summarizes the wavelength and the pertaining energy of different types of irradiation. For oxidation purposes, UV-C radiation is the most often in use, although in the literature could be found the application of other types of UV radiation [68, 73-75]. The most common wavelength of UV-C radiation is 254 nm that could be achieved by low-pressure vapor mercury lamp invented by Hewitt at 1901 [35]. The primary usage of UV radiation in the past was for disinfection, but with the development of reaction mechanism, UV radiation nowadays found the usage for oxidation purpose as well. At the room temperature, most molecules reside in their lowest-energy electronic state, i.e. "ground state". When molecules were exposed to UV radiation, they transferred to state with higher energy, i.e. "excited sate". The energy difference between the "ground" and the "excited" state depends on absorbed radiation $h\nu$, where h and ν are Planck constant and the frequency of absorbed radiation, respectively. The molecule in the "excited state" has very short lifetime (10^{-9} do 10^{-8} s), after which it returns to "ground state" by one of several mechanisms (fluorescence, phosphorescence...) or decomposes to yield a different molecule, resulting with photochemical reaction. Simple mechanism of direct photolysis is given below [35]:

$$M + h\nu \xrightarrow{k_a} M* \qquad (11)$$

$$M* \xrightarrow{k_b} M \qquad (12)$$

$$M* \xrightarrow{k_c} produkti \qquad (13)$$

UV radiation is always in use in combination with some powerful oxidant or photocatalyst. The efficiency of its separate use depends on the lot of limitation such as: (a) water solution should be treated in a way to achieve the highest possible UV transmission, i.e. turbidity should be as low as possible, (b) too high concentration of OH radicals could inhibit mineralization reaction of organic contaminant present in water, (c) water solution should be free of heavy metals and oils, and (d) the costs of UV radiation are higher than those for dark Fenton type processes. However, several authors investigated in their comparative studies the efficiency of UV radiation for the treatment of colored wastewater containing reactive dyes. Bali et al. [76] obtained only negligible decolorization, 2 %, of C.I. Reactive Black, initial concentration 100 mgL^{-1}, after two hours treatment by UV photolysis. Furthermore, Behnajady et al. [77] achieved after a one-hour treatment only 18.5 % of color removal with initial dye concentration of 30 mgL^{-1}. In our previous studies, in the case of C.I. Reactive Red 45 with initial concentration of 80 mgL^{-1} again poor color removal, 30 %, and the mineralization degree of 3 % of TOC removal were achieved [78]. Treating model dye solutions with lower initial dye concentrations, 20 mgL^{-1}, significantly higher color removals of 87 and 98 % were achieved, whilst only 17 and 6 % of dye solutions mineralized [52, 79]. Such low decolorization in the case of higher dye concentration could be the fact that although UV radiation itself has ability to destroy organic molecules [35], the efficiency of direct photolysis of organic dyes decay has been proved to be difficult and it depends on the dye's reactivity and photosensitivity, and more important, solution transmission is very low in the case of higher dye concentration [80]. Moreover, most of the commercially used dyes are usually designed to be light resistant.

Photochemical Processes

These processes combine UV radiation and the addition of some powerful oxidant, H_2O_2 and/or O_3, in the system. The application of UV/H_2O_2 process for water purification is widely investigated AOP due to the fact that involves the simplest way of OH radical generation, through direct photolysis of H_2O_2. For its successful application it is necessary to determine the type and the initial concentration of pollutants in water, as well as the presence of other organic and inorganic compounds, so-called "scavengers", that could inhibit or even stop the treatment process. This process is in use for the removal of micro- and macro-pollutants from drinking water, for the treatment of organic toxic compounds present in lower concentration in ground waters, for the treatment of smaller volume of highly recalcitrant pollutants in order to achieve their detoxification and faster degradation, for the control of exhaust gases in the case of volatile organic compounds (VOCs)... [81]. Quantity of energy that emits by direct photolysis of H_2O_2 under UV radiation is very high, and theoretically two OH radicals could be generated per absorbed energy quant, equation (8) [82]. In practice, the highest quantum yield for generation of OH radicals is 0.5 mol of H_2O_2 per Einstein [35, 83]. That value is called the basic energy of direct photolysis of H_2O_2. By using higher energy than basic, formed OH radicals can be recombined and in that way H_2O_2 is producing. According to the literature [84, 85], besides reaction (8), other important reactions should be also taken into account in UV/H_2O_2 process:

$$H_2O_2 \Leftrightarrow HO_2^- + H^+ \tag{14}$$

$$HO\bullet + H_2O_2 \rightarrow HO_2\bullet + H_2O \tag{15}$$

$$HO\bullet + HO_2^- \rightarrow HO_2\bullet + HO^- \tag{16}$$

$$HO_2\bullet + HO\bullet \rightarrow H_2O + O_2 \tag{17}$$

Above reactions present "scavenger" mechanism of H_2O_2 (14) or OH radicals (15-17) negatively influencing the overall process efficiency. Important process parameters of UV/H_2O_2, besides UV lamp characteristics and reactor performances, are pH of solution and the initial concentration of H_2O_2. One of limitations of UV/H_2O_2 process that should be taken into consideration is the presence of iron and potassium salts in treated water resulting with reduction of UV radiation, and that could be avoided by adjusting pH to the value where those salts can precipitate. Further limitation is related to the large quantities of suspended particles resulting with increased turbidity and that can be solved by the filtration as the pretreatment of such wastewaters. Besides H_2O_2, ozone is also very used as oxidant in photochemical processes. Moreover, ozone is even better oxidant than H_2O_2 due to the significantly higher value of molar absorption coefficient at 254 nm, typical value for UV-C radiation (3300 $M^{-1}cm^{-1}$>>>18.6 $M^{-1}cm^{-1}$). Also, the rate of ozone photolysis is almost 1000 times higher than that of

H_2O_2 [86]. UV/O_3 process is based on the fact that by the decomposition of ozone under UV radiation two OH radicals are generated which rather form H_2O_2 than react with organic matter present in water, shown by equation (18):

$$O_3 + H_2O \xrightarrow{h\upsilon} H_2O_2 + O_2 \qquad (18)$$

Furthermore, such formed H_2O_2 decomposes under UV radiation to OH radicals that react with organic matter present in water, equation (8). This oxidative system contains three components that could generate OH radicals and/or directly oxidize organic matter: UV radiation, H_2O_2 and O_3. Accordingly, there are several ways to generate OH radicals: (a) photolysis of formed H_2O_2 or (b) reaction between formed H_2O_2 and ozone, so-called peroxone reaction. Also, there are several mechanisms for the degradation of organic pollutants in water: (a) direct photolysis, (b) OH radical attack generated from different sources and (c) direct ozone attack [87]. Process efficiency is limited by concentration of ozone and other factors mentioned above in the case of UV/H_2O_2 process. There is also combination of these two binary systems (UV/H_2O_2 and UV/O_3) the $UV/O_3/H_2O_2$ process. This process could enable to achieve the complete mineralization of organics present in water due to the enhanced generation of OH radicals throughout added H_2O_2 in the system [36, 78].

In the literature could be found many studies concerning UV/H_2O_2 process and colored wastewaters, but application of UV/O_3 and $UV/O_3/H_2O_2$ processes are not so widely investigated, particularly for the treatment of wastewater containing reactive dyes. Mohey El-Dein et al. [88] achieved complete bleaching of C.I. Reactive Black 5 with initial dye concentration of 500 mgL^{-1} at pH 7 and 140 mM H_2O_2 solution after 150 minutes of treatment, while after 240 minutes mineralization reached almost 90 % of removed ecological parameter DOC (dissolved organic carbon). In our previous paper, time required for complete mineralization of C.I. Reactive Red 45 (80 mgL^{-1}) by UV/H_2O_2 process was 600 minutes, while complete bleaching was accomplished after 60 minutes of treatment [78]. Muruganandham and Swaminathan [89] achieved between 93.5 and 61.0 % of decolorization, whilst 62.9 and 24.6 % of aromatic products degraded after 150 minutes, applying UV/H_2O_2 process for the treatment of C.I. Reactive Orange 4 with various initial dye concentrations (0.1-0.5 mM). Bali et al. [76] decolorized C.I. Reactive Black (100 mgL^{-1}) model solution completely by UV/H_2O_2 process, pH 7 and initial H_2O_2 of 100 mM, after 60 minutes. In our previous study [79] where UV/H_2O_2 process was applied for the treatment of C.I. Reactive Blue 137 model solution with 20 and 80 mgL^{-1} initial dye concentrations, complete decolorization was achieved after 10 and 50 minutes, respectively. Colonna et al. [90] reported in their study of decolorization of several dyes, including one reactive, by UV/H_2O_2 process complete bleaching of C.I. Reactive Blue 19 after 110 minutes and that ratio [H_2O_2]/[dye] plays important role in the oxidation efficiency of applied system. That was also confirmed by Ince [91] and he named that ratio "effective H_2O_2 level". Shu and Chang [92] reported significantly faster decolorization of studied dyes by UV/O_3 than by UV/H_2O_2. Similar results are obtained in our previous study [78], where significantly faster decolorization and correspondingly higher mineralization of C.I. Reactive Red 45 was obtained by UV/O_3 and UV/O_3/H_2O_2 than by UV/H_2O_2 process. It was established that mineralization by UV/O_3/H_2O_2 is four time faster than that by UV/H_2O_2 process. As the important process parameter of UV/O_3 and UV/O_3/H_2O_2 processes is pointed out pH of solution and it was established that optimal pH is at neutral or weak basic conditions [78, 93].

Photocatalytic Processes

Heterogeneous photocatalysis is based on the photonic excitation of a solid, which renders it more complex. The term photocatalysis may designate several phenomena that involve photons and catalyst, while this part of the chapter will consider only semiconductor photocatalysis. Photocatalytic activity of TiO_2 is based on its semiconductor properties. Radiation by photons, which have higher transfer energy, of such semiconductor leads to generation of electron-hole pairs [94]:

$$TiO_2 \xrightarrow{h\upsilon} h_{vb}^+ + e_{cb}^- \qquad (19)$$

Holes in valence band (h_{vb}^+) are strong oxidants, while electrons in conductance band (e_{cb}^-) act as reductants. Holes in valence band reacts with H_2O or hydroxyl ions on the surface and producing OH radicals:

$$H_2O + h_{vb}^+ \rightarrow HO\bullet + H^+ \qquad (20)$$

$$h_{vb}^+ + HO^- \rightarrow HO\bullet \qquad (21)$$

Electrons react with dissolved oxygen producing superoxid radical ion, $O_2\bullet^-$, or its protonated form, perhydroxyl radical, $HO_2\bullet$, equations (22) and (23):

$$O_2 + e_{cb}^- \rightarrow O_2\bullet^- \qquad (22)$$

$$O_2\bullet^- + 2H^+ \Leftrightarrow 2HO_2\bullet \qquad (23)$$

In water, two $HO_2\bullet$ can be recombined if their concentration allow them to react significantly yielding H_2O_2 and O_2. It follows photocatalytic reduction of H_2O_2 by scavenging an electron from the conduction band where OH radicals are generated [95]:

$$H_2O_2 + e_{cb}^- \rightarrow HO\bullet + HO^- \qquad (24)$$

The efficiency of TiO_2 photocatalysis could be improved by the addition of H_2O_2, but the addition of optimal H_2O_2 dosage should be taken into account. When it is in excess, it decrease process effectiveness. Photocatalysts, i.e. semiconductors, are comprised of microcrystalline or nanocrystalline particles and they are used in a form of thin film or as powder dispersion [96, 97]. Besides TiO_2 as the mostly known, used and studied photocatalyst, the usage of alternative photocatalysts, such as ZnO, CdS, SnO_2, is investigated as well [71, 94-101]. One of the main operating parameter of photocatalytic process is the dosage of photocatalyst.

In the studies concerning the application of photocatalytic processes for reactive dye wastewater degradation, TiO_2 is mostly used photocatalyst. Bizani et al. [102] investigated the efficiency of UV/TiO_2 process for the degradation of two commercial reactive dyes (initial concentrations 50 mgL^{-1}) and achieved the complete decolorization between 30 and 120 minutes of treatment depending on dye molecular structure and the type of TiO_2 photocatalyst. Goncalves et al. [97] investigated optimal pH range for reactive dye decolorization by UV/TiO_2 process and they reported the highest rate of C.I. Reactive Orange 4 decolorization at strong basic condition. Muruganandham and Swaminathan [103] investigated the application of other photocatalysts such as SnO_2, ZnO, CdS and Fe_2O_3 for decolorization of C.I. Reactive Yellow, and they reported that SnO_2, CdS and Fe_2O_3 have negligible activity on RY14 decolorization and degradation. In our previous paper [104], two effective photocatalysts, TiO_2 and ZnO, are compared in the case of C.I. Reactive Red 45 model solution decolorization and mineralization. The optimal conditions for both processes are established and it was found out that there are no significant differences between these two photocatalysts for decolorization of model solution at established optimal conditions for each process. Similar findings are reported in the study Akyol et al. [105] where complete color removal of Remazol Red was obtained after 35 minutes treatment by UV/TiO_2 and UV/ZnO processes.

OZONE-BASED PROCESSES

Ozone is main component in many oxidation processes assembled under the term ozonation processes. In these processes ozone is applied either alone (O_3 process) or with the addition of oxidant, e.g. H_2O_2 (O_3/H_2O_2 process), UV radiation (explained in above subchapter), catalyst, activated carbon, ultrasound etc. Ozone is inorganic molecule constituted by three atoms of oxygen. It is present in nature in upper atmosphere in the form of stratospheric layer around the earth, and it is formed by the photolysis of diatomic oxygen and further recombination of atomic and diatomic oxygen, shown by equations (25) and (26) [35]:

$$O_2 \xrightarrow{h\upsilon} 2O\bullet \tag{25}$$

$$O\bullet + O_2 \rightarrow O_3 \tag{26}$$

Ozone can be generated artificially in ozone generators. There is two ways of generating ozone by ozone generators: (a) the cleavage of oxygen molecules under the influence of strong electrical field and (b) the same mechanism like in nature, the photolysis of oxygen, but induced artificially. The mechanism of both ways can be described by above reactions (25) and (26). Ozone was discovered in 1840, while its molecular structure as triatomic oxygen was established 30 years latter. The first usage of ozone was reported at the half of 19th century where ozone was used as disinfectant in many water treatment plants and hospitals. Therefore, at 1856 the first usage of ozone as disinfectant for operating rooms in hospital over Europe was reported [106, 107]. First usage of ozone for the purification of drinking water in Monaco and Germany was performed at 1860 [107, 108]. Ozone application as oxidant for water purification was retained through 20th century, and its significant increase was noticed at 1970s when the production of trihalomethanes and other organohalogenated hydrocarbons were identified during the water treatment by chlorine [35]. Ozone is very reactive reagent with redox potential of 2.07 V (table 2), either in liquid or gas phase. That high reactivity owes to electronic configuration; ozone can be presented as a hybrid in four different molecular resonance structures that give ozone characteristics of an electrophilic, dipolar or even nucleophilic agent. Hence, ozone molecule could react with organic compounds under two different mechanisms: direct and indirect. Direct mechanism involves organic compound degradation by molecular ozone, occurring at acidic pH range. The main reactions involved in direct mechanism are the reaction of addition to the unsaturated part of the hydrocarbon molecule and electron transfer. Rather high oxidation/reduction potential enables ozone to react with many organic, but also inorganic compounds [35].

$$O_3 + 2H^- + 2e^- \rightarrow O_2 + H_2O \tag{27}$$

Degradation of organics by indirect ozone mechanism occurs at basic conditions and involves generation of OH radicals and their further reaction with present organic compounds in water. Hydroxyl radicals are generated by the reaction of ozone with hydroxyl ions present in water [109]:

$$O_3 + H_2O \xrightarrow{OH^-} HO\bullet + O_2 + HO_2\bullet \tag{28}$$

Mechanism of ozone decomposition in water depends on the presence of chemical species that can initiate, promote and/or inhibit its decomposition. The most accepted ozone decomposition mechanism is given in Figure 3.

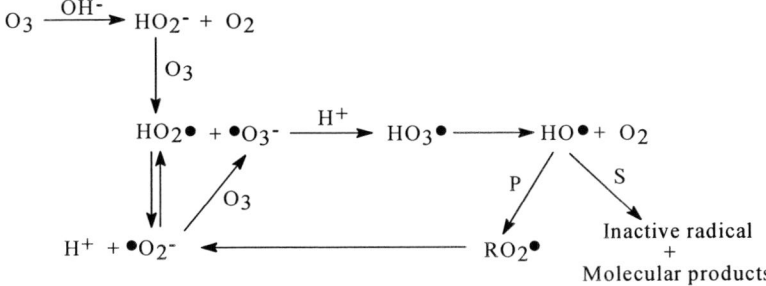

Figure 3. Scheme of ozone decomposition mechanism in water. P = promoter (e.g. ozone, methanol), S = scavenger or inhibitor (e.g. *t*-butanol, carbonate ion), I = initiator (e.g. hydroxyl ion, perhydroxyl ion) (adapted by Beltran [35]).

According to the presented mechanism of ozone decomposition in water, it can be seen that ozonation can classified in AOPs, when the degradation of organics occurs by indirect mechanism, and in classical chemical treatment methods, when direct mechanism is dominant in the degradation of organics in water. One of the limiting factors of ozonation influencing overall process effectiveness presents the efficiency of ozone mass transfer from gas to liquid phase. One of possible solutions gave Mathews et al. [110] by the application of *in situ* ozone generator with porous electrode. They obtained 90 % of C.I. Reactive Blue 19 decolorization in two times shorter treatment period than by the conventional ozone generator. Besides O_3 process, either direct or indirect, one of widely investigated process from the group of ozone-based processes is O_3/H_2O_2, so-called peroxone process or perozonation [111]. Efficiency of that process is based on the fact that the presence of H_2O_2 could enhance ozone reactivity in bulk phase. Moreover, some authors consider degradation of organics by direct mechanism negligible in the case of O_3/H_2O_2 process [111, 112]. Hydroxyl radicals are generated by reactions of ozone and perhydroxyl ion, HO_2^-, which is present in water by the dissociation of H_2O_2. The simple overall mechanism of OH radical generation in peroxone process is shown by equation (29) [112]:

$$H_2O_2 + O_3 \rightarrow 2HO\bullet + 3O_2 \tag{29}$$

Degradation kinetic of organic pollutants by O_3 process, either direct or indirect mechanism, is reported as first order or "pseudo"-first order kinetic, while in the case of peroxone process, degradation kinetics follows second order due to the predominant degradation mechanism by OH radicals.

Many investigations are available in the case of ozonation as treatment methods for colored wastewaters containing reactive dyes. Chu and Ma [30] investigated direct and indirect O_3 process for the degradation of several dyes including three reactive. They reported dye decolorization by indirect mechanism significantly faster than that by direct mechanism. Wu and Wang [113] reported the complete decolorization of C.I. Reactive Black 5 (0.5 gL^{-1}) in 40 minutes of treatment by O_3 process with rate of ozone generation 26.1 $mgL^{-1}min^{-1}$, and their results revealed that rate of ozone transfer increased with increase in dye concentration, applied ozone dose and temperature. Peralta-Zamora et al. [114] obtained very fast complete decolorization of C.I. Reactive Blue 19, in <5 minutes, but degradation of dye monitored in UV region, i.e. degradation of aromatics formed after the initial cleavage of dye chromophores, was significantly slower. Furthermore, Arslan-Alaton et al. [115] investigated the application of O_3

process for the treatment of simulated dyebath effluent containing several different reactive dyes at different pH region: acidic, neutral and basic. Results of their study showed that decolorization of simulated dyebath is very fast at all three pH regions, yielding with 100 % color removal in couple of minutes, and that degradation of aromatics, monitored at 280 nm, was somewhat slower than decolorization, and that overall extent of aromatics degradation depends on operating pH value. Mineralization of solution was also monitored, and significantly lower extents of partial mineralization were achieved than those of decolorization and aromatic degradation. They obtained similar findings in previous study too [116]. Similar large difference between decolorization and mineralization extents were obtained in our previous comparative study of O_3, O_3/zeolite (NH_4ZSM5 or HY type) and O_3/H_2O_2 processes for degradation of C.I. Reactive Blue 137 model solution [117]. Model dye solution was completely decolorized at all initial dye concentrations (20-150 mgL^{-1}), while only partial mineralization was obtained in all cases. The somewhat higher mineralization extents were obtained by O_3/H_2O_2 than by O_3 process. Zeolites did not affected decolorization rate, only slightly due to changing of pH values of solution, but somewhat higher TOC removals were obtained in cases with zeolite addition than without. Explanation could be found in the adsorption of aromatic by-products with appropriate molecular structure to penetrate in zeolite pores. Besides the improvement of O_3 process by H_2O_2 addition, some authors combined O_3 and ultrasound processes in order to improve degradation of reactive dye solution [118]. From the presented results it can be seen that decolorization of colored wastewater can be easily achieved by ozonation processes, but dye degradation in order to get complete mineralization is very hardly to achieve.

HIGH VOLTAGE ELECTRICAL DISCHARGE PROCESSES

Basic principles of processes including electrical discharge, so-called corona discharge, in liquid and gas phase, are that chemical and physical processes could be induced by strong electrical field, and therefore these processes could be consider as electrical AOPs. It is proved that by applying electrical discharge in liquid phase intensive UV irradiation and various active species (•OH, •H, •O, HO_2^-, $O_2^{•-}$, H_2O_2, etc.) are considered to be produced [39, 40, 119-124]. By the presence of oxygen in the system, ozone and its ozone-related radical species can be formed. This process can be effective for the treatment of biological microorganisms and dissolved chemicals in liquid phase [125]. Application of the processes based on strong electric field in water or organic solution is investigated in the past due to importance in electrical transmission processes and they practical usage in biology, chemistry and electrochemistry. Recently, the application of reactors involving electrical discharge in liquid or above the surface, for the treatment of drinking and wastewaters is widely investigated. Nowadays, these reactors are in use, but only in laboratory scale. There are many types of corona reactors according to the configurations of their high voltage and ground electrodes. The most common reactor type has both electrodes, high voltage and ground, in liquid phase [40]. Generally, corona is type of electrical discharge that is formed when DC, AC or pulsed electric potential is applied between two non-uniform electrodes where a dielectric medium is placed between the electrodes. Corona onset begins when the discharge initiates in the dielectric medium. That manifests as thin violet layer which surrounds the high voltage electrode in liquid phase. That layer contains of very short streamers. Initial voltage of corona discharge depends on the type of applied electrodes and physical properties of dielectric material between the two electrodes. As the corona discharge voltage increases, the streamers become larger and corona discharge is formed. Further increasing of the voltage leads to the sparkover, i.e. the production of direct current channel that bridges two electrodes [119, 125]. As it was mentioned before, pulsed corona discharge is applied in gas and liquid phase. In gas phase corona, the mostly used electrode geometry is "plate-to-plane" and "wire-to-cylinder", while in liquid phase the most studied geometries are "point-to-plane" and "plate-to-plane" (Figure 4) [40].

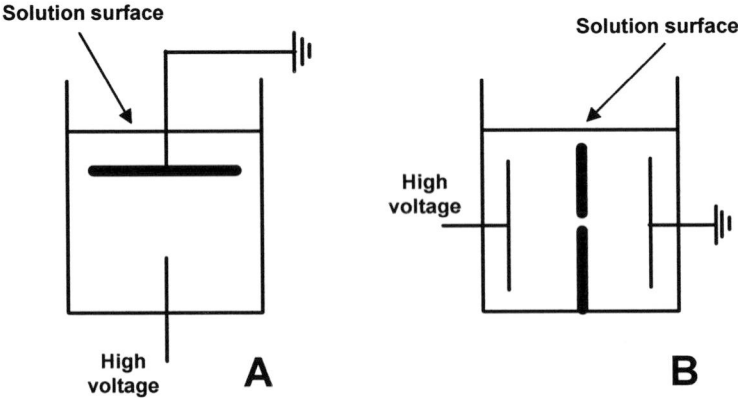

Figure 4. Most studied electrode geometry in liquid corona discharge reactors: (A) point-to-plane and (B) plate-to-plane.

Pulsed corona discharge produce repetitive (60 Hz), fast raising high voltage pulses with short lifetime (μs). Application of short high voltage pulses (200-1000 ns) consider presence of non-thermal conditions, that means electrons in plasma having a higher mean energy ($T_e \gg 1$ eV), than the other constituents in liquid ($T_{H2O} < 0.1$ eV) [121, 126]. On this way, the energy loss for the migration of ions, chemical species with less mobility than electrons and which have no influence to generation of radicals, is minimized as high as possible [120]. Among all radical species generated in corona when electrons with high energy collide with water molecules, equations (30) and (31), OH radicals present main species responsible for degradation of organics.

$$H_2O + e^-* \rightarrow HO\bullet + H\bullet + e^- \qquad (30)$$

$$H_2O + e^-* \rightarrow H_2O^+ + 2e^-* \qquad (31)$$

$$H_2O + H_2O^+ \rightarrow H_3O^+ + HO\bullet \qquad (32)$$

Such formed radical species could react between themselves resulting with the production of H_2O_2, O_2, or H_2.

$$H\bullet + H\bullet \rightarrow H_2 \qquad (33)$$

$$HO\bullet + HO\bullet \rightarrow H_2O_2 \qquad (34)$$

$$HO\bullet + H\bullet \rightarrow H_2O \qquad (35)$$

Degradation of organics could be performed by: (a) direct reactions with highly reactive species (OH radicals), (b) indirect reactions over radicals formed from stabile molecules (H_2O_2), and (c) direct reactions with stabile molecules.

Hybrid Corona Reactors

These types of corona reactors involve simultaneous application of non-thermal plasma above liquid surface and direct high voltage electrical discharge in liquid phase. That is achieved by the combination of different electrode configuration and/or by changing electrode geometry of standard corona reactors which use "point-to-plane" geometry. There are two hybrid corona reactors, so-called hybrid-series and hybrid-parallel (Figure 5). Hybrid-series involves "point-to-plane" geometry and configuration of electrodes where high voltage electrode is in liquid phase, and ground electrode is placed in gas phase just above the liquid solution. That enables the formation of conventional streamer like corona discharge in liquid and the formation of intensive plasma channels near to liquid surface. Hybrid-parallel corona reactor involves electrode geometries "point-to-plane" in liquid phase and "plane-to-plate" in gas phase. That reactor configuration combines mostly used electrode geometries in liquid and gas phase, i.e. on that way is combined classical corona discharge in liquid and gas.

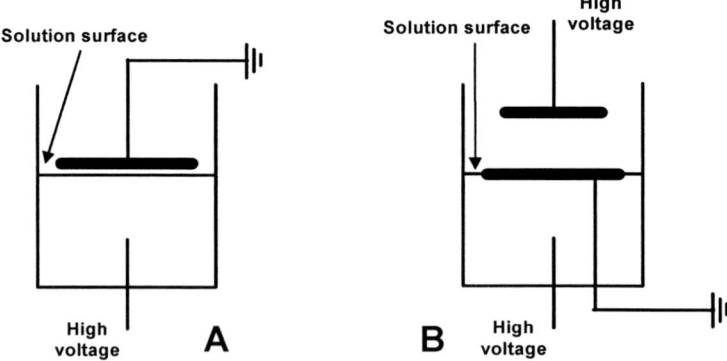

Figure 5. Configuration of hybrid corona reactors: (A) series and (B) parallel.

Both hybrid reactor types are very interested due to improvements concerning standard liquid corona type reactor (Figure 4). It is reported that hybrid reactors, besides other earlier mentioned radical, ionic and molecular species, include generation of ozone in gas phase and its transfer to liquid phase, improving degradation of organics in liquid [127]. Besides explained corona discharge, there are other types of discharges in water such as glow, barrier, glidarc, but they are not discussed in this subchapter.

There are not too many studies directed to the investigation of corona discharge reactors application for the treatment of colored wastewater, particularly containing reactive dyes. Tri Sugiarto et al. [128] investigated the application of corona discharge reactor, "plane-to-plate" geometry, for decolorization of three dyes (initial dye concentrations of 10 mgL^{-1}). They reported complete decolorization at neutral pH after 120 minutes treatment and dependence of decolorization rate on pH value. They obtained the highest decolorization rate at basic conditions. Sunka et al. [129] investigated new type of corona discharge reactor, "pin hole" geometry, for decolorization of C.I. Reactive Blue 137 (50 mgL^{-1}). In first case where only corona discharge was used for decolorization of reactive dye solution, very low color removal was achieved after 30 minutes of treatment, while the addition of Fe ions in the system improved color removal and in 5 minutes of treatment the complete color removal was obtained. Similar findings were obtained in study by Loncaric Bozic et al. [130] treating C.I. Reactive Blue 49 in the same reactor type. One of our earliest paper related to corona discharge dealt with the application of corona discharge reactor with "point-to-plane" electrode geometry for decolorization of model solution of four different dyes (20 mgL^{-1}), including two of reactive type [131]. Decolorization extents ranged between 17-25.6 % after 60 minutes of treatment, while by the addition of Fe ions in the system, complete decolorization was achieved after 30 minutes. Our two new papers related to corona are directed to the application of hybrid corona reactors for decolorization of reactive dyes [132, 133]. C.I. Reactive Blue 137 (20 mgL^{-1}) was decolorized up to 72.9 and 85.0 % regarding reactor type. Both papers also investigated the influence of synthetic zeolites on decolorization by corona discharge and it was found that zeolites could improve color removal in hybrid-parallel corona reactor. Decolorization effectiveness was decreased by increasing initial dye concentration, resulting with color removal in the case of 80 mgL^{-1} initial dye concentrations between 34 and 48 % regarding dye structure, type of reactor, the addition of zeolites or not, and type of added zeolite. From the presented results of corona discharge processes for decolorization of colored wastewaters it can be seen that these type of processes could be efficient, but only when dye loading in wastewater is not so high.

OTHER AOPS

This part of subchapter contains less detailed review of ambient AOPs that are not studied in our laboratory, such as: ultrasound or ultrasonic irradiation, radiolysis of water and electrochemical processes.

Ultrasound

The term sonochemistry describes all chemical processes in which ultrasound irradiation is involved. The main premise of ultrasound process as wastewater treatment method is generation of free radicals, such as hydrogen and hydroxyl, upon the action of ultrasonic waves in liquid. The applied frequency ranges from 20-40 kHz; predominantly used in large scale applications, so-called "power ultrasound", up to 1 MHz; frequency used in so-called "diagnostic ultrasound" [134]. Ultrasound produces the chemical effects through several different physical mechanisms and the most important nonlinear acoustic process for sonochemistry is cavitation. Cavitation can be generated when large pressure differentials are applied in flowing liquid (hydrodynamical cavitation), or by means of an electrochemical transducer, piezoelectrical or magnetostrictive, in contact with the fluid (acoustical cavitation) [135]. Electrohydraulic cavitation in liquid causes bubbles. These bubbles start very small, and grow in diameter and implode under the periodic variations of the pressure field of the ultrasonic waves. The rapid implosion of the eventually instable gas bubbles causes adiabatic heating of the bubble vapor phase. It is assumed that creates drastic local conditions: temperatures of about 5500 °C inside the bubble and 2100 °C in the liquid that surrounds the cavity, and pressure up to 100 MPa within the collapsing cavity. Such conditions are limited to a very small region and the heat produced during cavitation is dissipated very quickly (heating and cooling rates greater than 109 K/s). As a result, the surrounding liquid remains at the ambient temperature. Organic compounds are highly degraded in such an environment and inorganic compounds can be oxidized or reduced. Water irradiation using ultrasound causes decomposition of the water molecules into extremely reactive radicals HO• and H•, shown by equation (36) [119, 136].

$$H_2O \xrightarrow{ultrasound} H\bullet + \bullet OH \qquad (36)$$

Several authors investigated ultrasound as treatment process for colored wastewater degradation. Tezcanli-Guyer and Ince [137] investigated ultrasound, 520 kHz, for the treatment of four organic dyes including two of reactive type (initial concentrations 20-60 mgL^{-1}), and reported decolorization after two hours treatment between 81-99 % depending on dye structure, while degradation of formed aromatics after four hours treatment ranged between 11-93 %. Rehorek et al. [138] applied power ultrasound, 850 kHz, at different input powers for the treatment of several organic dyes. They obtained complete decolorization of all studied dyes in 1-15 hours of treatment regarding input power of ultrasound, and they reported that ultrasound is able to degrade organic dyes to non-toxic products.

Water Radiolysis

Irradiation of dilute aqueous solutions by high energy ionizing radiation, ranged from keV to MeV, results with the excitation and the ionization of water molecules leading to the production of radical species. The types of radiation may be classified by the source of production: (a) the decay of radioactive nuclei (α-, β- and γ-radiation), (b) beams of accelerated charged particles (electrons, protons and deutrones) and (c) short-wavelength electromagnetic radiation (X rays). OH radicals and hydrated electrons, e_{aq}^-, can be considered as predominant species formed by water radiolysis. Besides these two species, water radiolysis produces also hydrogen atoms, hydrogen peroxide (predominant specie in the presence of O_2), hydroxyl ions, etc. The mostly investigated radiation sources for degradation of organics are γ-radiation and electron beam [119, 139-141]. Dajka et al. [140] reported the role of e_{aq}^- in decolorization of C.I. Reactive Black. Zhang et al. [141] investigated γ-radiation for decolorization of three different dyes and reported that, depending on dye structure, different gamma irradiation dose are required to achieve the complete decolorization.

Electrochemical Processes

These processes can occur by a direct electron transfer reaction to (reduction) or from (oxidation) the present organic pollutant, or by a chemical reaction of the pollutant with previously electrogenerated species. The mechanism is generally viewed as a direct anodic oxidation of organic pollutant involving its reduction by direct electron transfer from organic molecule to the electrode to form a radical cation that readily deprotonates, equation (37):

$$RH \xrightarrow{electrolysis} R\bullet + H^+ + e^- \qquad (37)$$

Nowadays, various types of electrodes are in use: Pt/IrO_2, Ti/Pt, SnO_2/Ti, $Ti/RuTiO_2$, etc [119, 142]. Catanho et al. [143] used $Ti/RuTiO_2$ electrode for decolorization of C.I. Reactive Red 198 and reported 75 % color removal after 2 hours treatment. Awad and Abo Galwa [144] investigated different conductive electrolytes for decolorization of colored wastewater and obtained the highest decolorization in the case of NaCl. Carneiro et al. [145] investigated different types of electrochemical processes for degradation of C.I. Reactive Blue 4 and reported the as the best process $Ti/SnO_2/SbO_x/RuO_2$ electrode in Na_2SO_4 0.2 M at pH 2.2 and potential of +2.4 V, where complete decolorization and 60 % mineralization were obtained.

Chapter 3

STUDY OF C.I. REACTIVE BLUE 137 WASTEWATER; TREATMENT BY SEVERAL AOPS

After a review of available and currently studied AOPs, two types of processes were chosen for our research directed to degradation of C.I. Reactive Blue 137 as pollutant in model wastewater. Due to the very high effectiveness and fastness of Fenton type processes, but also considering secondary treatment of solution due to iron ions and counter ions, Fenton type process using iron powder, Fe^0/H_2O_2, was one of studied treatment processes. It was concluded that the assistance of UV light to Fe^0/H_2O_2 could offer significant benefit to process effectiveness due to the achievement of complete mineralization, as it was mentioned above in "Fenton type processes" subchapter. Another group of chosen processes include those combining ozone and UV light, UV/O_3 and $UV/O_3/H_2O_2$, due to the conductance at neutral or at weak basic conditions that excludes after-treatment solution conditioning like in the case of Fenton type processes, and also due to constant ozone supply enabling to constant OH radical generation. The influence of synthetic zeolites on chosen AOPs was also investigated in the framework of this research. Synthetic zeolites were shown positive effect when added to some of AOPs for the treatment of colored wastewater in our previous studies [104, 117, 132, 133, 146, 147]. Generally, synthetic zeolites, are three-dimensional, microporous, crystalline solids with well defined structures that contain aluminum, silica, and oxygen in a regular framework. Zeolites have void pore space that can host cations, water or other molecules. They have the ability to act as catalysts for chemical reactions which take place within the internal cavities. Furthermore, synthetic zeolites may act as ion exchangers, because the loosely-bound nature of the extra-framework metal ions allows the exchange of other types of metals when present in aqueous solution. Shape-selective properties of zeolites are also the basis for their use in molecular adsorption [148-150]

OBJECTIVES

The aim of the study was to find the optimal treatment process for colored wastewater containing reactive dye C.I. Reactive Blue 137. The comparative study involved several processes, Fe^0/H_2O_2, $UV/Fe^0/H_2O_2$, UV/O_3 and $UV/O_3/H_2O_2$, and it was performed through an experimental research of decolorization and mineralization of a reactive azo dye C.I. Reactive Blue 137, as a model wastewater pollutant. Applied processes were optimized according to their process parameters, Fe^0 dosages, initial H_2O_2 dosages and initial pH values. The influence of initial organic dye concentration as well as the addition of solid particles, synthetic zeolites, on the process effectiveness was also investigated. Studied AOPs were evaluated on the basis of their eco-effectiveness, by the means of color (A_{610}), aromaticity (A_{280}), TOC and AOX value decrease, and their cost-effectiveness as well.

MATERIALS AND METHODS

Synthetic wastewater containing organic model pollutant was prepared using two different concentrations, 20 and 80 mgL^{-1}, of the commercial dye Cibacron Marine P2R-01, C.I. Reactive Blue 137 (RB137) obtained from Ciba-Geigy, and deionezed water; pH 7, conductivity less than 1 µS cm^{-1}. Other chemicals (H_2O_2, 30 %; sulfuric acid, >95 %; and sodium hydroxide, p.a.) were supplied by Kemika, Croatia.

Series of experiments were conducted in order to find an optimal Fe^0/H_2O_2 ratio. At the beginning of the experiment, pH was adjusted at 3 using 25 % sulphuric acid, which was followed by the addition of iron powder and H_2O_2. Iron powder dosages were 28 and 56 mgL^{-1} that corresponded to the concentrations of 0.5 and 1.0 mM, respectively. Concentrations of hydrogen peroxide were varied to give molar ratios 1 : 5, 1 : 10, 1 : 20, 1 : 30, 1 : 40, and 1 : 50. Reaction mixture (V=250 mL) was continuously stirred at room temperature in an open batch system with magnetic stirring bar and was treated for two hours, while dye concentration and TOC values were measured at the end of the each experiment to established decolorization and mineralization extents.

All further experiments were carried out in a 0.8 L batch glass annular water-jacketed photoreactor, used in our previous studies [48, 52, 59, 78, 79, 87]. The value of incident photon flux at 254 nm, 7.36×10^{-6} Einstein s^{-1}, was calculated on the basis of hydrogen peroxide actinometry experiments [151]. In experiments that required the addition of iron powder, Fe^0/H_2O_2 and $UV/Fe^0/H_2O_2$, optimized dosages of iron powder and H_2O_2 were added accordingly to the results of experiments explained in previous paragraph (Fe^0/H_2O_2 optimization study part). In experiments where UV irradiation was required, $UV/Fe^0/H_2O_2$, UV/O_3 and $UV/H_2O_2/O_3$, the UV lamp was switched on. In the experiments involving ozone, UV/O_3 and $UV/H_2O_2/O_3$, ozone was introduced into the reactor. Ozone was generated from pure oxygen, >99.9 %, by introducing it into the ozone generator, MIC Systems Inc., Valdosta, Georgia, USA to produce ozone. The outlet gas mixture from the ozone generator, i.e. mixture of ozone and oxygen, was fed into the reactor through a sintered glass plate diffuser located in the bottom half of the reactor. The inlet gas flow into the ozone generator was fixed at 0.15 Lmin^{-1}. The rate of ozone generation in reactor inlet stream, 7.9×10^{-3} gmin^{-1}, was determined iodometrically [152]. In experiments of UV/O_3 and $UV/H_2O_2/O_3$ processes optimization, initial pH values and hydrogen peroxide concentration ranged from 3 to 11, and 0 to 10 mM, respectively, until the achievement of maximal mineralization extent. Initial pH was adjusted with the addition of NaOH or H_2SO_4. Added quantities of H_2SO_4, NaOH and H_2O_2 as an additional oxidizing agent, were negligible in comparison to the total volume of treated reaction mixture. The experiments were conducted by adjusting one variable, while others were held constant.

The total volume of the treated solution was 0.5 L in all cases, while the mixing of the solution was provided by both magnetic stirring and peristaltic pump at a flow rate of 0.1 Lmin^{-1}. Experiments were carried out at 25 (+/- 0.2) °C. The duration of each experiment was one hour. Samples were taken periodically from the reactor (2, 5, 10, 15, 20, 30, 40, 50 and 60 min) and thereafter immediately analyzed. In experiments where the influence of synthetic zeolites on process effectiveness was investigated, at the beginning of experiments different amounts of zeolites were added (0.4 gL^{-1} of NH4ZSM5 in the case of Fe^0/H_2O_2 process and 1.0 gL^{-1} in the cases of UV/O_3 and $UV/H_2O_2/O_3$ processes), accordingly to our previous studies [117, 146]. In these experiments, samples were centrifuged before the analysis in order to remove solid particles of zeolites from the solution. All experiments were repeated at least two times to give reproducibility of the experiments within 5 %, and averages are reported.

The initial pH values were measured by handylab pH/LF portable pH/conductivity-meter, Schott Instruments GmbH, Mainz, Germany. The decolorization and degradation of RB137 solution was monitored by a Perkin Elmer Lambda EZ 201 UV/VIS spectrophotometer, USA, by scanning complete UV and VIS spectra. The extent of RB137 mineralization was determined on the basis of measurements of total organic carbon (TOC), using TOC analyzer; TOC-V_{CPN} 5000 A, Shimadzu, Japan, and absorbable organic halides (AOX), performed by Organic Halide Analyzer, DX-2000, Dohrmann.

RESULTS AND DISCUSSION

Fenton Type Processes

First group of investigated AOPs were addressed to Fenton type process. As it was mentioned earlier, the oxidation power of Fenton type processes, particularly those performed in the dark; strongly depends on process parameters such as iron catalyst concentration, concentration of H_2O_2, pH, temperature, and treatment time. Therefore, it was necessary to perform laboratory investigations, so-called "jar tests", of those parameters in order to find optimal for chosen system Fenton type process/organic pollutant [65]. In the research "jar tests" were directed to find optimal concentration of iron powder and optimal Fe^0/H_2O_2 ratio, while other process parameters were in the range reported as optimal in the literature: pH 3, room temperature, and a one-hour treatment time [44, 45, 49, 51, 52, 54, 66]. The results of Fe^0/H_2O_2 process parameters for the treatment of RB137 model solution with initial concentration of 20 mgL^{-1} are summarized in Figure 6. It can be seen that through complete investigated range of Fe^0/H_2O_2 ratio at both iron powder dosages rather high decolorization of RB137 colored wastewater, >95 %, was achieved. Obtained decolorization extents were below visibility limit, <1 mgL^{-1} of remained dye in the solution [2]. During the same treatment period, only partial mineralization was obtained in all cases, with the highest 40.9 % of TOC removal in the case of c(Fe^0)=1.0 mM and 1 : 20 Fe^0/H_2O_2 ratio. It can be concluded that decolorization of RB137 was not significantly influenced by investigated process parameters, but at the same time, mineralization showed different results. Almost the same system behavior could be observed at both investigated iron powder dosages and the same Fe^0/H_2O_2 ratio. Mineralization of RB137 increased with increasing the Fe^0/H_2O_2 ratio, but only to the certain point. That can be explained with the increasing of OH radical concentration in the bulk, resulting with an increase of mineralized part of RB137. According to the results of TOC removal, further increase over obtained optimal ratio showed negative effect to the process effectiveness. That can be explained with the fact that H_2O_2 when is in excess, act as OH radical scavenger, resulting with the lowering of RB137 mineralization [44, 45, 47-49, 65]. Besides that effect, the creation of an inert oxidative film on the surface of iron powder at higher concentration of H_2O_2 should also be taken into account. That inert film could disable the leaching of Fe^{2+} ions from iron powder surface in the bulk, as well as the occurrence of Fenton reaction on the iron powder surface [54].

After determining the optimal iron powder dosage and Fe^0/H_2O_2 ratio, the kinetic of RB137 degradation with Fe^0/H_2O_2 and $UV/Fe^0/H_2O_2$ processes was investigated. In Figure 7. the results of RB137 degradation with both processes are presented. By monitoring complete UV/VIS spectra, it was tried to get an overview on the degradation of complete dye molecular structure.

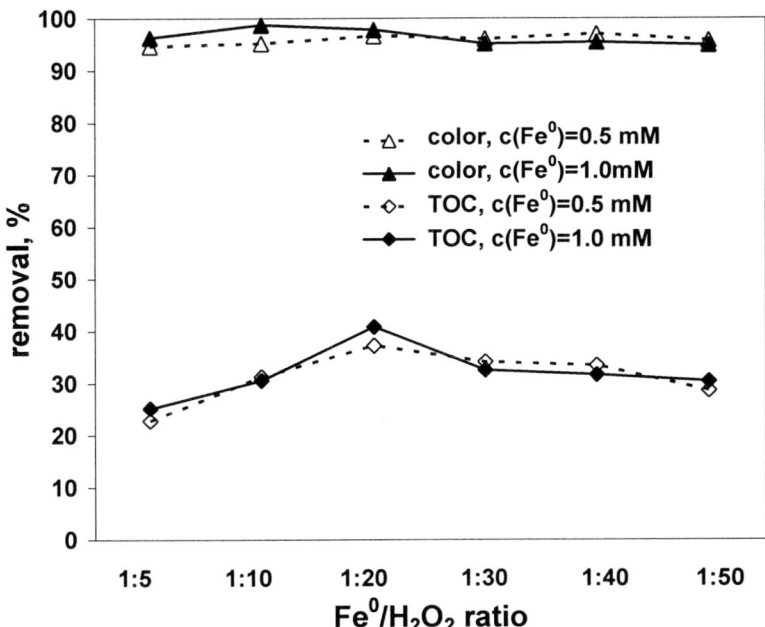

Figure 6. Influence of iron powder dosages and Fe^0/H_2O_2 ratio on color and TOC removal in Fenton "like" process for degradation of RB137 at pH 3.

By full black line the chromatogram of RB137 through complete UV/VIS spectra was presented. Three characteristic peaks can be observed. First one absorbing in the VIS region between 480-680 nm with maximum wavelength at

610 nm corresponds to RB137 chromophore. Next two peaks absorb in UV region, first between 260-350 nm with maximum wavelength at 280 nm corresponding to the absorbance of aromatic part of dye molecule and second between 200-260 with maximum wavelength at 230 nm corresponding to triazine ring in dye molecular structure [45, 72, 153, 154]. Comparing degradation of dye chromophores, i.e. the disappearance of peak absorbing at λ_{max}=610 nm, it can be observed that decolorization is much slower with Fe^0/H_2O_2 than $UV/Fe^0/H_2O_2$ process. Therefore, complete decolorization of RB137 by $UV/Fe^0/H_2O_2$ process was achieved after 5 minutes of treatment, while in the case of Fe^0/H_2O_2 process, the complete bleaching, i.e. zero absorbance of RB137 model solution at 610 nm was observed barely after 20 minutes treatment process. The band of RB137 in UV region with two peaks at 230 and 280 nm was interfered through almost complete treatment time by absorbance of added H_2O_2 [35, 155]. It can be observed that the monitoring of dye degradation in UV region, as well as degradation of aromatic structures and triazine ring was partially disabled due to the interference by H_2O_2. But, it can be seen that new peak, produced by the interference of triazine ring peak with the absorbance of added H_2O_2, is lowering much faster in the case of $UV/Fe^0/H_2O_2$ than Fe^0/H_2O_2 process. That might be due to the much faster consumption of added H_2O_2, but also due to much faster degradation of triazine rings, by process combined with UV light. That is in accordance with the literature where is reported that $UV/Fe^0/H_2O_2$ process consumed H_2O_2 much faster than Fe^0/H_2O_2 process [48]. Furthermore, triazine rings have very high molar absorption coefficient at 230 nm, ≈ 40000 $M^{-1}cm^{-1}$, that makes them very unstable under UV-C irradiation [72]. Degradation of aromatic structures at λ_{max}=280 nm is also much faster by $UV/Fe^0/H_2O_2$ than by Fe^0/H_2O_2 process. Although that peak was also interfered by the addition of H_2O_2, in the case of $UV/Fe^0/H_2O_2$ process a significant decrease of that peak during the treatment period from 2.-20. minute could be observed, meaning that aromatic structures in RB137 molecules were degraded. After 30. minute of treatment time, the appearance and the constant increasing of new peak in UV region with maximum wavelength at 290 nm can be observed. That peak could correspond to the formation of organic acids as products of degradation of aromatic compounds in model solution that absorb in UV region at 220 and 290 nm [71, 153]. That is also confirmed by the increasing of peak at 220 nm in treatment period between 40. and 60. minute.

Study of C.I. Reactive Blue 137 Wastewater; Treatment by Several AOPs 37

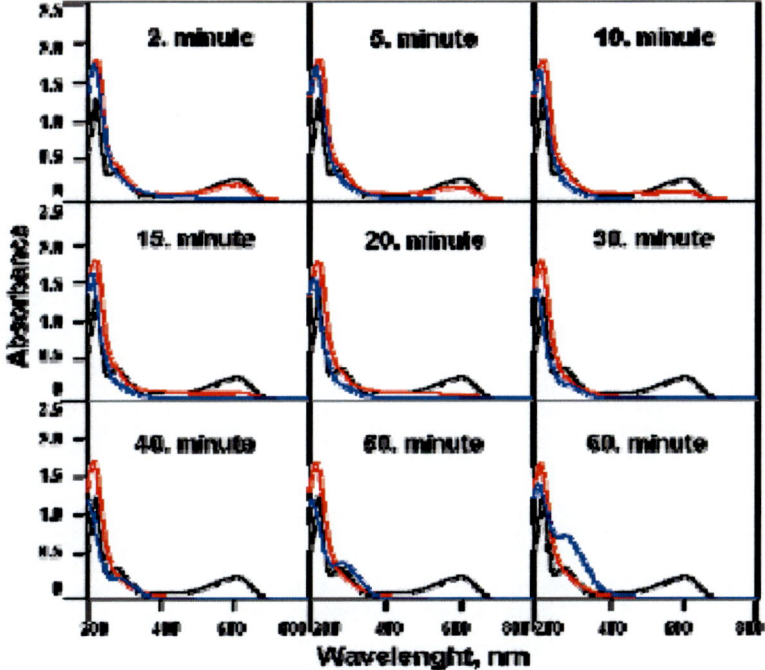

Figure 7. Degradation of RB137 with initial concentration 20 mgL^{-1} (_____) by Fe^0/H$_2$O$_2$ (_____) and UV/Fe^0/H$_2$O$_2$ (_____) processes monitoring through complete UV/VIS spectra.

The higher efficiency of UV/Fe^0/H$_2$O$_2$ process for degradation of RB137 colored wastewater in comparison Fe^0/H$_2$O$_2$ process at the same process parameters can be also observed from the results of TOC removal during the treatment by those two processes (Figure 8). In the first treatment period, until 15. minute, that difference is not so pronounced. But after 15. minute, in the case of Fe^0/H$_2$O$_2$ process strong inhibition in mineralization of RB137 model solution can be observed. On the other hand, the inhibition of UV/Fe^0/H$_2$O$_2$ process occurred after 40. minute of process. Inhibition and possible complete stopping of dark Fenton type process could be due to the complete consumption of H$_2$O$_2$, or due to the breaking of Fenton catalytic cycle by the formation of Fe^{3+}-complexes, mostly carboxylates [64]. Inhibition of photo-Fenton type processes could be caused by complete consumption of H$_2$O$_2$ and in that case degradation of organics will be continued only by their direct photolysis. Further reason for the inhibition of photo-Fenton type process might be due to the formation of organics acids resistant to degradation by OH radicals, such as acetic and oxalic [37], which can be degraded under UV light but with lower rate than other organic content by OH radicals. In this case it is more likely that the inhibition of both processes was not caused by the complete consumption of H_2O_2, due to the results of rather high remained H_2O_2 concentration in bulk, >20 %, obtained in our previous study at the same process conditions, but with other organic pollutant [48].

Hence, it can be assumed that the inhibition of dark process, Fe^0/H_2O_2, occurred due to the formation of Fe^{3+}-complexes resulting with a lack of OH radicals in bulk produced through bulk Fenton reaction, while the mineralization efficiency of $UV/Fe^0/H_2O_2$ process was inhibited by the formation of large amounts of OH radical attack resistant organics acids. That is in accordance with the observation from Figure 7, where the appearance of new peaks occurred after 30. minute of treatment process.

The possible improvement of dye degradation efficiency of dark Fenton type process was tried to get by the addition of synthetic zeolite NH_4ZSM5. In our previous study [146], that type of zeolite showed positive effect on TOC removal by Fe^0/H_2O_2 process when added in lower dosages, <0.5 gL^{-1}. In table 4 the comparison of Fe^0/H_2O_2, $Fe^0/H_2O_2/NH_4ZSM5$ and $UV/Fe^0/H_2O_2$ processes concerning the final decolorization, degradation of chloro-triazine rings and mineralization of RB137 model solution (20 mgL^{-1}) after a one-hour treatment is presented. It can be observed that all three compared processes showed very high decolorization efficiency, >99 % color removal in all cases. It should be noted that so-called "blank-tests" were performed in order to investigate adsorption of dye molecules on zeolites or possible penetration in zeolites pores. No changes of dye spectra monitored through complete UV/VIS region were obtained, indicating no dye adsorption by NH_4ZSM5 zeolite. Comparing mineralization efficiency of organic content by all three investigated processes, it can be observed that partial mineralization extents were obtained in all cases.

Table 4. Degradation efficiency of RB137 model solution (20 mgL^{-1}) by Fe^0/H_2O_2, $Fe^0/H_2O_2/NH_4ZSM5$ and $UV/Fe^0/H_2O_2$ processes

process	Fe^0/H_2O_2	$Fe^0/H_2O_2/NH_4ZSM5$	$UV/Fe^0/H_2O_2$
color removal, %	100	99.2	100
TOC removal, %	40.8	51.5	83.5
AOX removal, %	42.2	42.5	90.5

The increase of TOC removal for 10.3 % by $Fe^0/H_2O_2/NH_4ZSM5$ process in comparison to Fe^0/H_2O_2 could be contributed to the addition of solid particles. But, it is more likely that synthetic zeolite acted as adsorbents for molecule of appropriate size to be adsorbed/penetrated on/in zeolite. No effect of zeolite addition to other monitored parameters, color and AOX removal, and the findings of our previous study where scanned IR spectrograms of zeolites after the treatment with $Fe^0/H_2O_2/NH_4ZSM5$ showed new peaks which corresponded to possible dye degradation by-products [146], indicating to above mentioned assumption. Moreover, Kawai and Tsutsumi [157] reported, and the same was confirmed in our previous studies [59, 158], that NH_4ZSM5 zeolite could adsorb phenol. If phenol could be adsorbed, it can be assumed that other mono-substituted benzenes could also be removed from the aqueous solution by adsorption on/in NH_4ZSM5 zeolite. Furthermore, Feng et al. [156] proposed degradation mechanism of reactive dye with similar structure like RB137. They reported that after the initial cleavage of azo bonds, in the bulk are remained mono- and di-substituted benzenes and naphtholes and triazine rings. Moreover, Stylidi et al. [71] detected phenol and other mono-substituted benzenes as by-products of azo dye degradation by OH radicals. In the favor of adsorption of only mono-substituted benzenes speaks also the fact that AOX removal was almost the same in the case with and without zeolite addition, indicating inability of NH_4ZSM5 zeolite for adsorption of triazine rings. The highest removal of all parameters; color, TOC and AOX was obtained when Fe^0/H_2O_2 process is combined with UV irradiation, resulting with 105 and 115 % improvement in TOC and AOX removal efficiency.

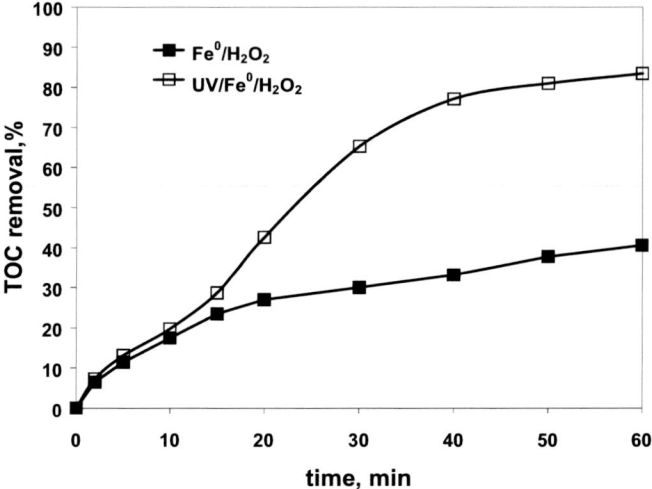

Figure 8. Mineralization of RB137 with initial concentration 20 mgL^{-1} by Fe^0/H_2O_2 and $UV/Fe^0/H_2O_2$ processes monitoring decrease of TOC ecological parameter.

The efficiency of Fe^0/H_2O_2 and $UV/Fe^0/H_2O_2$ processes was also tested on colored wastewater containing 80 mgL^{-1} of RB137 (Figure 9).

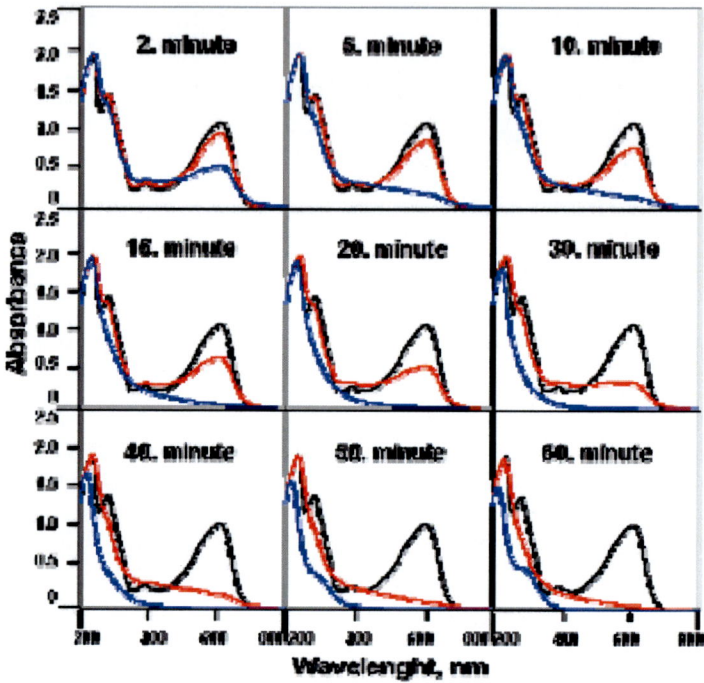

Figure 9. Degradation of RB137 with initial concentration 80 mgL^{-1} (_____) by Fe^0/H_2O_2 (_____) and $UV/Fe^0/H_2O_2$ (_____) processes monitoring through complete UV/VIS spectra.

Again, three specific peaks can be observed: dye cromophore peak at 610 nm, aromatic peak at 280 nm and triazine ring peak at 230 nm. The peaks in UV region were also interfered by absorbance of added H_2O_2, but not in such high manner like in the case with the initial dye concentration of 20 mgL^{-1}. $UV/Fe^0/H_2O_2$ process again showed rather high decolorization efficiency of colored wastewater; the complete bleaching was achieved between 20. and 30. minute of treatment, while in the case of Fe^0/H_2O_2 process complete bleaching was not achieved even after 60 minutes treatment. From Figure 9 a rather fast degradation of aromatics, i.e. disappearance of peak at 280 nm, can be observed even more clearly than in case of lower initial dye concentration due weaker interference of that peak by H_2O_2 absorbance. On the other hand, in the case of Fe^0/H_2O_2 process the lowering of that peak was much slower than in the case of

UV/Fe^0/H$_2$O$_2$ process. Also, an appearance of new peak with maximum wavelength at 290, corresponding to the formation of aliphatic by-products, was observed after 40. minute of treatment in the case UV/Fe^0/H$_2$O$_2$ process.

Mineralization of RB137 solution (80 mgL^{-1}) by those two processes is showed in Figure 10. Again, the pronounced difference in process effectiveness can be observed after 10. minute of treatment, where the inhibition of Fe^0/H$_2$O$_2$ process can be observed.

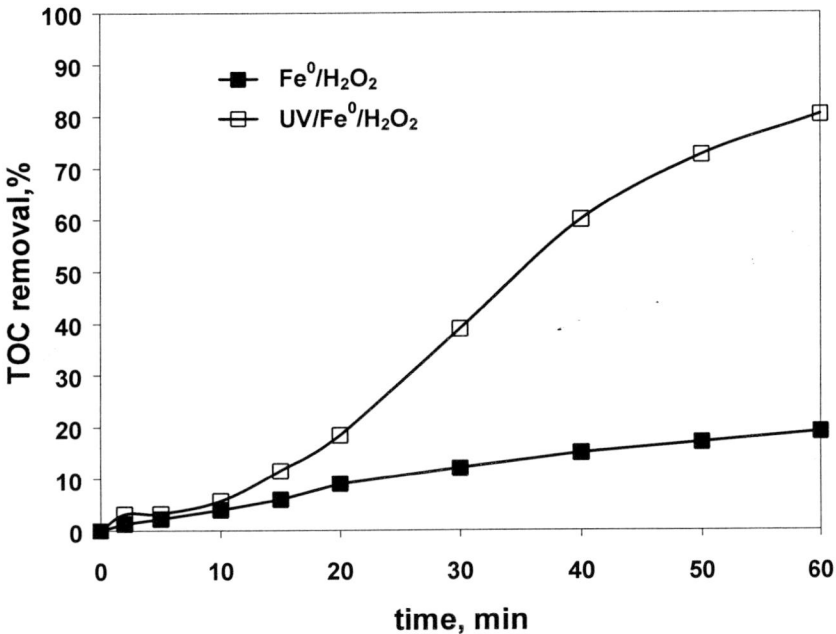

Figure 10. Mineralization of RB137 with initial concentration 80 mgL^{-1} by Fe^0/H$_2$O$_2$ and UV/Fe^0/H$_2$O$_2$ processes monitoring decrease of TOC ecological parameter.

In the case of UV/Fe^0/H$_2$O$_2$ process strong inhibition like in the case of lower initial dye concentration is not observed, but it is obvious that the rate of mineralization process is lowering by increasing of mineralization degree of model solution. It should be noted that final achieved mineralization extents; 19.1 and 80.3 % of TOC removal by Fe^0/H$_2$O$_2$ and UV/Fe^0/H$_2$O$_2$ process, respectively, are not four times lower than those obtained in the case of lower dye concentration, 40.8 and 83.5 %. Moreover, overall removed organic content expressed in mgC per L are much higher in the case of higher dye concentration: 4.9 >> 2.6 in the case of Fe^0/H$_2$O$_2$ and 20.4 >> 5.3 in the case of UV/Fe^0/H$_2$O$_2$

process. Such a large difference in the overall amount of mineralized dye might be due to the fact that besides OH radicals some other radical species, produced by the destruction of chromophoric parts and aromatic rings of the dye molecules and dye by-products, may be involved in further chain reactions of the dye degradation mechanism. Namely, with higher dye concentration, after initial cleavage of the chromophoric parts of the dye molecules, more radical species originated from chromophores and aromatic by-products destruction are present in the bulk and may participate in further dye degradation/mineralization. A similar effect is known from the literature. Chen and Pignatello [47] reported the formation and the involvement of organic radical species in phenol degradation mechanism by OH radicals. Furthermore, the simple mechanism proposed for the UV radiation chemistry is very similar. When UV radiation is incident upon the dye molecule, the radiation can be adsorbed, promoting the molecule to an excited state which can be involved in further degradation of dye molecules via radical mechanism [35, 159].

UV-Based Processes

Next group of investigated AOPs were photochemical processes: UV/O_3 and $UV/O_3/H_2O_2$. Like in Fenton type processes, the efficiency of UV-based processes strongly depends on process parameters such as: light intensity, reactor configuration, pH, type of oxidant, concentration of oxidant. Therefore, it was necessary to find optimal parameters for those which are changeable in our case; pH and initial H_2O_2 concentration (when added), while other process parameters, such as ozone feed concentration, light intensity, were held constant. In Figure 11 final color removals, TOC removals after 30 and 60 minutes treatment, and final pH values after the treatment of RB137 model wastewater (20 mgL^{-1}) by UV/O_3 process through initial pH range between 3 and 11 are presented. According to the complete color removal, achieved in <10 minutes of treatment, through entire pH investigated, it can be concluded that bleaching is not influenced by initial pH value. On the other hand, presented results of TOC removal after 30 and 60 minutes treatment were influenced by initial pH value, particularly at strong basic conditions.

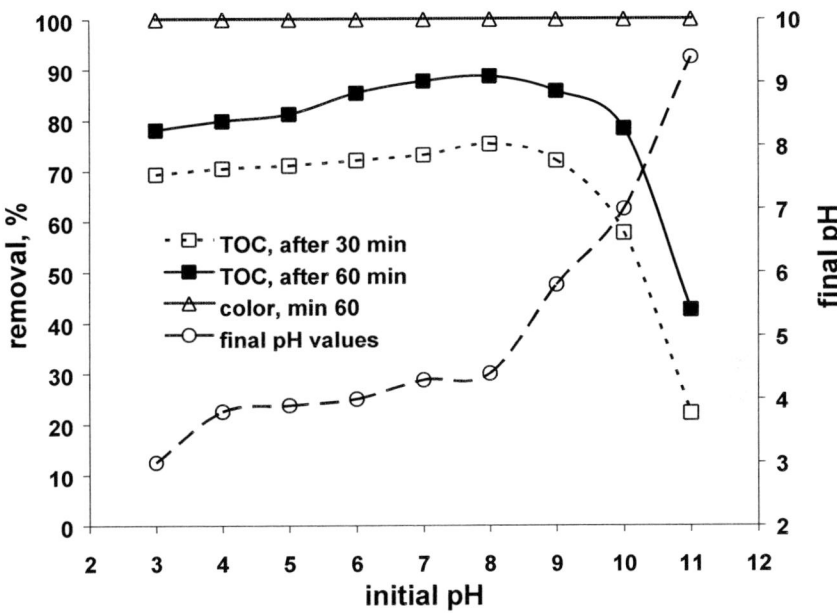

Figure 11. Influence of the initial pH value on the decolorization and mineralization efficiency, and final pH values after a one-hour treatment by UV/O$_3$ process.

The optimal pH area for the mineralization of RB137 is found to be between 6-9, i.e. at neutral and weak basic conditions, with the highest TOC removal, 88.7 %, achieved after 60 minutes at pH 8. It can be seen that the mineralization process is much faster in first 30 minutes of treatment process than in next 30 minutes. That can be attributed to the formation of large quantities of organic acids, including those resistant to OH radical degradation, in the last 30 minutes of treatment with significantly lower mineralization rate. The decrease in pH value after the treatment in comparison to initial value corresponds to the formation of acid by-products, mostly aliphatic [92].

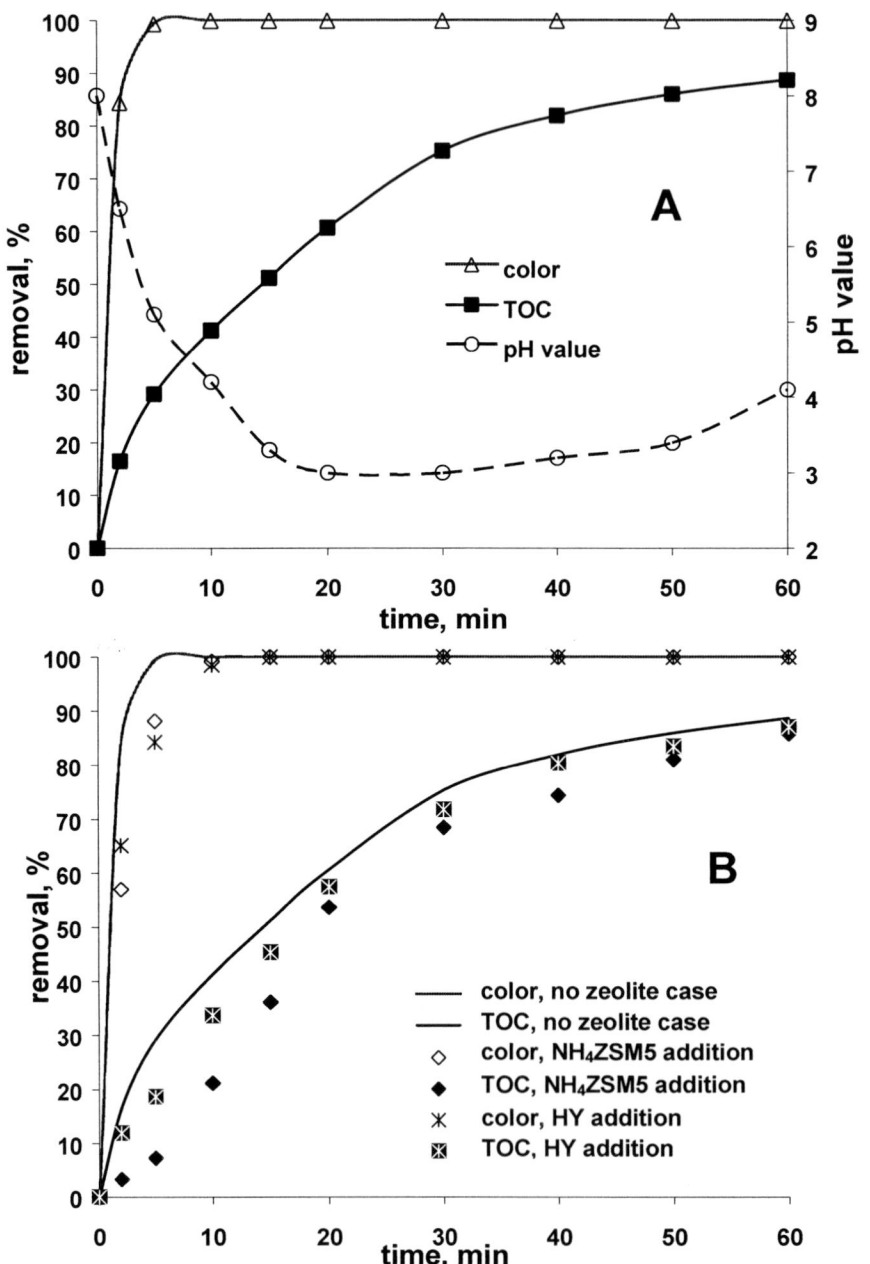

Figure 12. RB137 (20 mgL^{-1}) decolorization and mineralization efficiency as well as changes of pH values during the treatment (A), and influence of NH$_4$ZSM5 and HY zeolites on RB137 degradation efficiency (B) by UV/O$_3$ process (pH 8).

After optimizing UV/O$_3$ process regarding initial pH value, the kinetic of RB137 model solution (20 mgL^{-1}) decolorization and mineralization, as well changes of pH value during the treatment by UV/O$_3$ process (pH 8) were monitored (Figure 12.A). It can be seen that complete bleaching was achieved even after 5 minutes of treatment, while the rate of RB137 mineralization was decreased when treatment process progressed. That is particularly pronounced in last 30 minutes of treatment time. As it was mentioned above, the reason for such system behavior could be the formation of large quantities of organic acids, mostly aliphatic, as by-products after initial cleavage of azo bonds in dye molecules, and subsequent formation of aromatic by-products, and thereafter their degradation [71, 160]. Some of formed acids are recalcitrant to degradation through OH radial degradation mechanism [37]. However, they can be destroyed by UV irradiation, but slower than other organic content by OH radicals.

Additional reason for the slowing down of mineralization process could be found in the fact that mineralization efficiency of UV/O$_3$ process is slightly lower in strong acidic pH region than at neutral and weak basic pH (Figure 11). Formation of acid by-products is also confirmed by large drop in pH value. As treatment process progressed, pH value started slightly to increase, that can be attributed to the mineralization of acid by-products. In Figure 12.B are presented results of experiments with the addition of synthetic zeolites NH$_4$ZSM5 and HY to UV/O$_3$ process. These two types of zeolites were added to the oxidative system combining UV irradiation and ozone due to the observed positive effect of those zeolite types on decolorization of model wastewaters containing reactive dyes in hybrid-parallel corona reactor, which involves, besides classical corona discharge in liquid and gas phase, UV irradiation in liquid, as well as ozone formation in gas and its transfer to liquid phase [132, 133]. It should be noted, that like in the case of NH$_4$ZSM5 zeolite, series of experiments for the determining adsorption characteristics of HY zeolite for RB137 removal from aqueous solution were conducted showing no dye adsorption/penetration on/in zeolite. From Figure 12.B it can be seen that the addition of synthetic zeolites showed slight negative influence to the rate of RB137 decolorization and mineralization. That can be explained by the fact that efficiency of processes based on UV irradiation strongly depends on solution turbidity, i.e. solution transparency and clearness [35, 59].

The efficiency of UV/O$_3$ process (pH 8) for decolorization and mineralization of RB137 was investigated at higher dye loading of model wastewater, 80 mgL^{-1} (Figure 13). It can be seen that complete bleaching was achieved after 20 minutes of treatment, while final extent of partial mineralization after a one-hour treatment is 64.4% TOC removal. Trend of the curve that connects points of TOC removal extents measured during the treatment is almost linear, i.e. very different than that in the case of lower initial dye concentration. Although drop in pH value indicate the formation of acidic by-products, the mineralization of RB137 model solution was not slowed, indicating that the most of formed by-products are not resistant to OH radical attack. But, with further process progressing over 60 minutes, the slowing down of mineralization process could be expected, due to the fact that Hoigne and Bader [162] pointed out that organic acids with low reaction rate constants under UV irradiation and OH radical resistant, such as acetic and oxalic, always accumulate as final products when any type of reactive organic aqueous solutes are degraded by ozone or OH radicals in water. Hence, the overall oxidation of organic material to CO_2 and H_2O will be delayed whenever the oxidations lead to oxalic or acetic acid as intermediates, which take part at acids conditions. Again, like in the case of lower dye loading, the influence of synthetic zeolites to UV/O_3 process efficiency was investigated (Figure 13.B). It can be seen that zeolite addition caused slight delay in decolorization of RB137 model wastewater (80 mgL^{-1}). That can be explained by earlier mentioned increase in the solution turbidity, that decrease UV irradiation efficiency for ozone photolysis resulting with slower generation of OH radicals.

On the other hand, the addition of NH_4ZSM5 zeolite showed slight positive effect to TOC removal in treatment period after 30. minute, but that was not the case with HY zeolite. If addition of zeolites negatively influence decolorization, it can be assumed that negatively effects the mineralization too, due to the same reason; increase in system turbidity. But, such different system behavior concerning the addition of different type of zeolite could be explained with the adsorption/penetration of organic molecules formed as dye degradation by-products with appropriate size on/in zeolite NH_4ZSM5. Namely, these two zeolites are different according to pore size and structure, Si/Al ratio, surface area, etc. ZSM5 type zeolite has a pentasyl structure which is characterized by parallel channels with pores sizes 0.53-0.56 nm with crossed channels with pores sizes 0.51-0.55 nm, while Y zeolite has a basic cubic structure with pores sizes 0.74 nm [148-150]. As it was proposed above in the case of $Fe^0/H_2O_2/NH_4ZSM5$, NH_4ZSM5 zeolite showed ability to adsorb mono-substituted benzenes. According to the results presented in Figure 13.B, where TOC removal raised over that obtained in the non-zeolite case, it can be assumed that at particular treatment time, formed mono-substituted benzenes as by-products of dye degradation adsorbed on zeolites or even penetrated in zeolite pores. Such system behavior in process using ozone is already observed in our previous study [117].

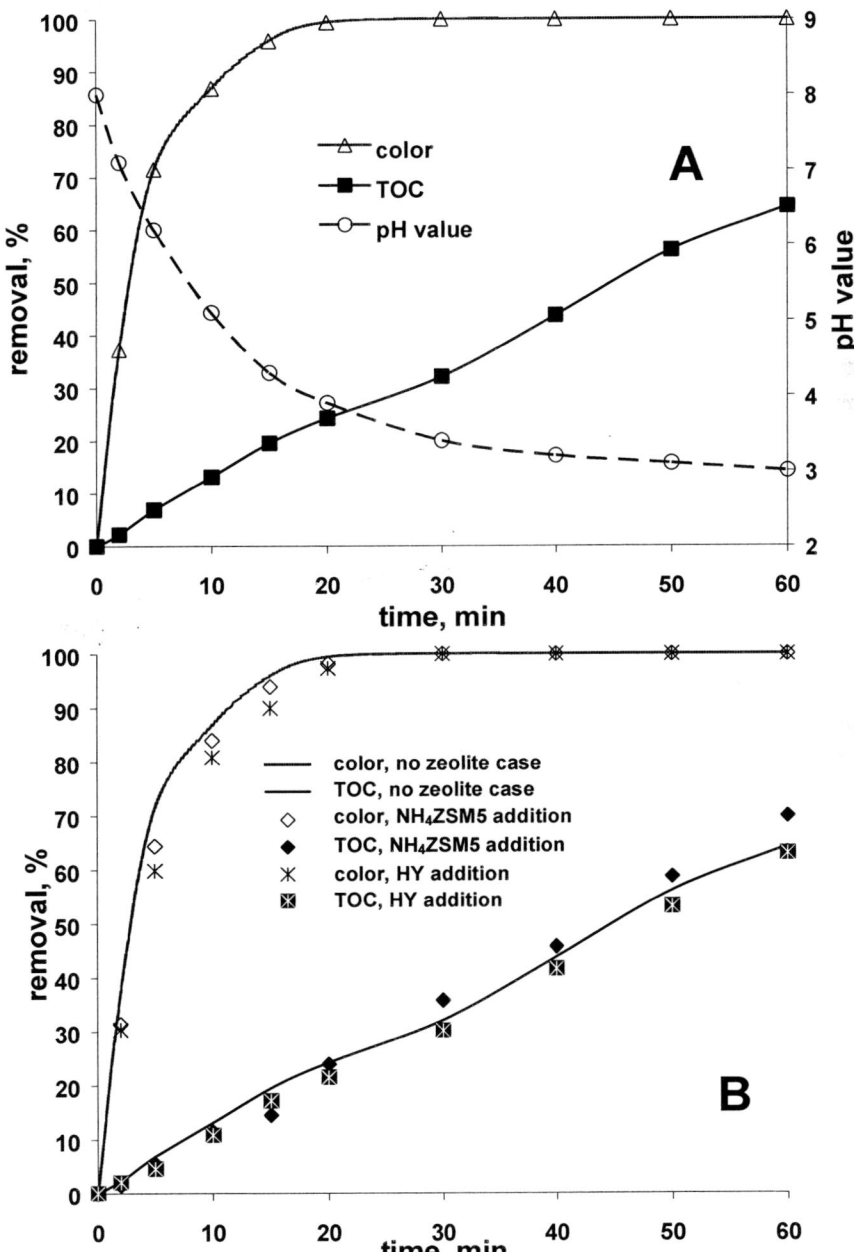

Figure 13. RB137 (80 mgL^{-1}) decolorization and mineralization efficiency as well as changes of pH values during the treatment (A), and influence of NH$_4$ZSM5 and HY zeolites on RB137 degradation efficiency (B) by UV/O$_3$ process (pH 8).

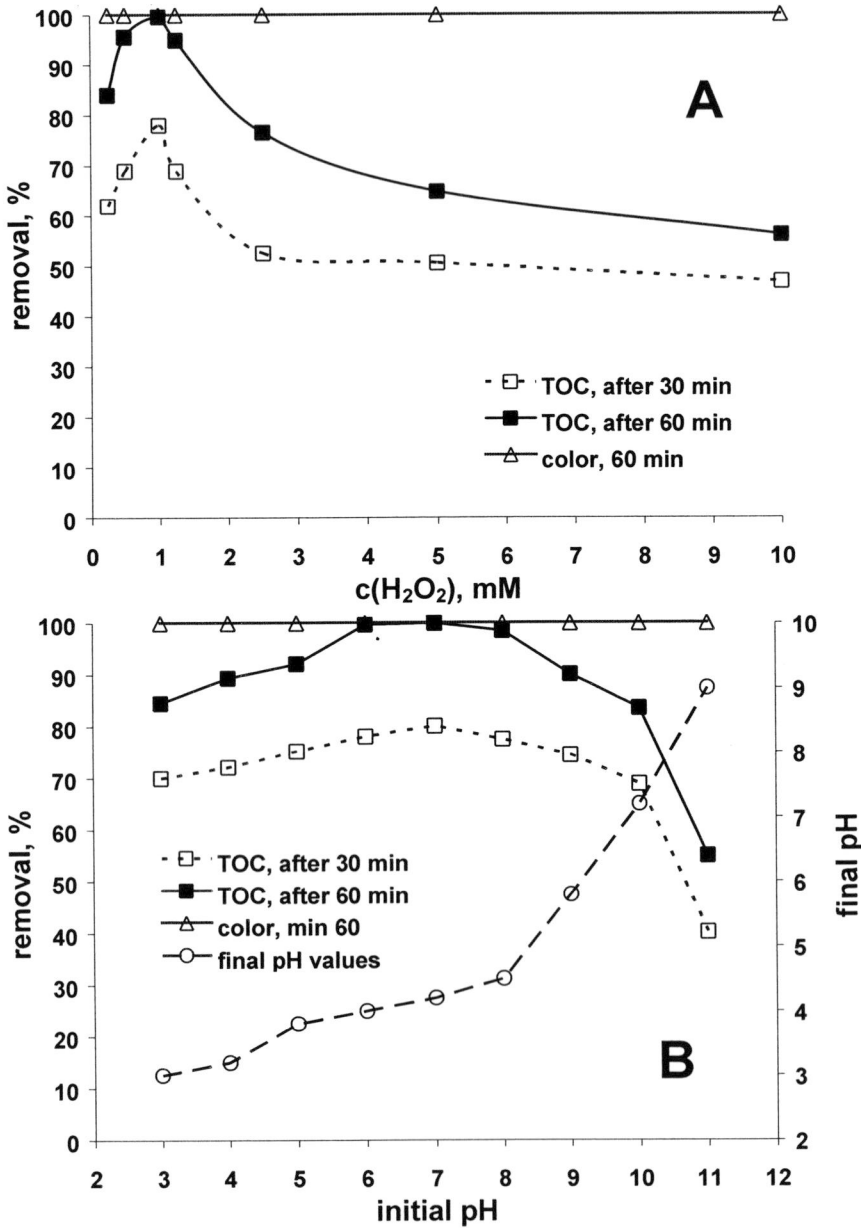

Figure 14. Influence of the initial H_2O_2 concentration (A), and initial pH value (B) on the decolorization and mineralization efficiency, and final pH values after a one-hour treatment by $UV/O_3/H_2O_2$ process.

In the same manner like UV/O$_3$ process, UV/O$_3$/H$_2$O$_2$ process was optimized regarding initial pH value, but also regarding to the other important changeable process parameters; initial H$_2$O$_2$ concentration. The results of optimization study are presented in Figure 14. Firstly, UV/O$_3$/H$_2$O$_2$ process was optimized concerning initial H$_2$O$_2$ concentration by changing that parameter and keeping others constant, including initial pH too, at zero pH charge of RB137 model wastewater (pH 6). It can be seen that the complete color removal was achieved through entire range of initial H$_2$O$_2$ investigated (0.25-10 mM). Moreover, that is achieved in very short treatment period, <10 minutes. Hence, it can be concluded that initial H$_2$O$_2$ concentration does not influence to much on decolorization efficiency of UV/O$_3$/H$_2$O$_2$ process. But, as it can be seen, initial H$_2$O$_2$ concentration strongly influence the mineralization efficiency. By increasing the initial H$_2$O$_2$ concentration, RB137 mineralization efficiency of UV/O$_3$/H$_2$O$_2$ process also increased, but only to the certain level, where further increase of H$_2$O$_2$ concentration negatively influences to the system effectiveness. Explanation for such system behavior is given as follows: slight enhancement of mineralization extents in the first part of the range of initial H$_2$O$_2$ concentration studied (0.25-1.0 mM), could be contributed to an increase of OH radical concentration in the bulk according to H$_2$O$_2$ photolysis mechanism, shown by equation (8). The decrease of RB137 degradation efficiency by UV/O$_3$/H$_2$O$_2$ process with further increasing of the initial H$_2$O$_2$ concentration could be the result of H$_2$O$_2$ bulk concentration higher than 1.0 mM. When hydrogen peroxide is in excess, it behaves as a scavenger, consequently lowering the hydroxyl radicals concentration and decreasing the overall mineralization efficiency of applied process [78, 87]. Similar system behavior was observed during the optimization of H$_2$O$_2$ concentration in Fenton type process (Figure 6). By comparing mineralization extents of RB137 model wastewater by UV/O$_3$ at pH 6 (Figure 11) and by UV/O$_3$/H$_2$O$_2$ at the same pH with optimal concentration of H$_2$O$_2$ (Figure 14.A), it can be seen that almost 20 % higher TOC removal was obtained by UV/O$_3$/H$_2$O$_2$ process.

In Figure 14.B the optimization of UV/O$_3$/H$_2$O$_2$ process regarding to the initial pH value is presented. That was performed in the way that pH values were changed from 3-11, while initial H$_2$O$_2$ concentration was held constant at 1.0 mM, found as optimal in the previous set of experiments (Figure 14.A). Again, complete bleaching was achieved in very short treatment period through complete range of pH investigated. According to mineralization extents achieved after 30. and 60. of minute treatment, the optimal pH region for degradation of RB137 by UV/O$_3$/H$_2$O$_2$ process was established. It can be observed similar system behavior like in the case of UV/O$_3$ process (Figure 11); a significant decrease in process effectiveness at strong basic conditions, and the highest mineralization efficiency at neutral and weak basic pH region. But it should be pointed out that the addition of H_2O_2 to UV/O_3 system improved mineralization efficiency through complete investigated pH range. Moreover, the complete mineralization was achieved after a one-hour treatment in the pH region between 6 and 8. Due to such system behavior, optimal pH was found out according to the mineralization extents obtained after 30 minutes treatment time. On that way, it was found the highest mineralization of RB137, 80.1 % TOC removal, by $UV/O_3/H_2O_2$ process ($c(H_2O_2)=1.0$ mM) at pH 7. It should be noted that the trend of curve that connects points of final pH values measured after a one-hour treatment by $UV/O_3/H_2O_2$ process (Figure 14.B) is very similar to that obtained by UV/O_3 (Figure 11).

Degradation kinetics of RB137 (20 mgL^{-1}) by $UV/O_3/H_2O_2$ process at optimal process conditions; pH 7 and $c(H_2O_2)=1.0$ mM, is showed in Figure 15.A. It can be seen that complete bleaching was achieved after 5 minutes of treatment. The same system behavior, like in the case of the treatment of RB137 (20 mgL^{-1}) by UV/O_3 process, slowing down of mineralization process with the progress of treatment process can be observed. However, complete mineralization was achieved. Decrease in the rate of mineralization is due to the earlier mentioned reason, production of large quantities of organic acids with lower rate of degradation. The drop in pH during the treatment time corresponds to the formation of acid by-products, while the achievement of complete mineralization resulted with reaching the initial pH 7 at the end of the process. Like in the case of UV/O_3/zeolite system for degradation of RB137 (20 mgL^{-1}) (Figure 12.B), the same system behavior could be observed in the case of $UV/O_3/H_2O_2$/zeolite (Figure 15.B). Moreover, negative influence of synthetic zeolites NH$_4$ZSM5 and HY addition is even more pronounced to RB137 decolorization and mineralization efficiency.

The $UV/O_3/H_2O_2$ process efficiency was tested for decolorization and mineralization of RB137 model wastewater with higher dye loading (Figure 16). In all three cases: non-zeolite (Figure 16.A), with the addition of NH$_4$ZSM5 and with addition of HY zeolite (Figure 16.B) the very similar system behavior like that showed in the case of UV/O_3 process (Figure 13) can be observed.

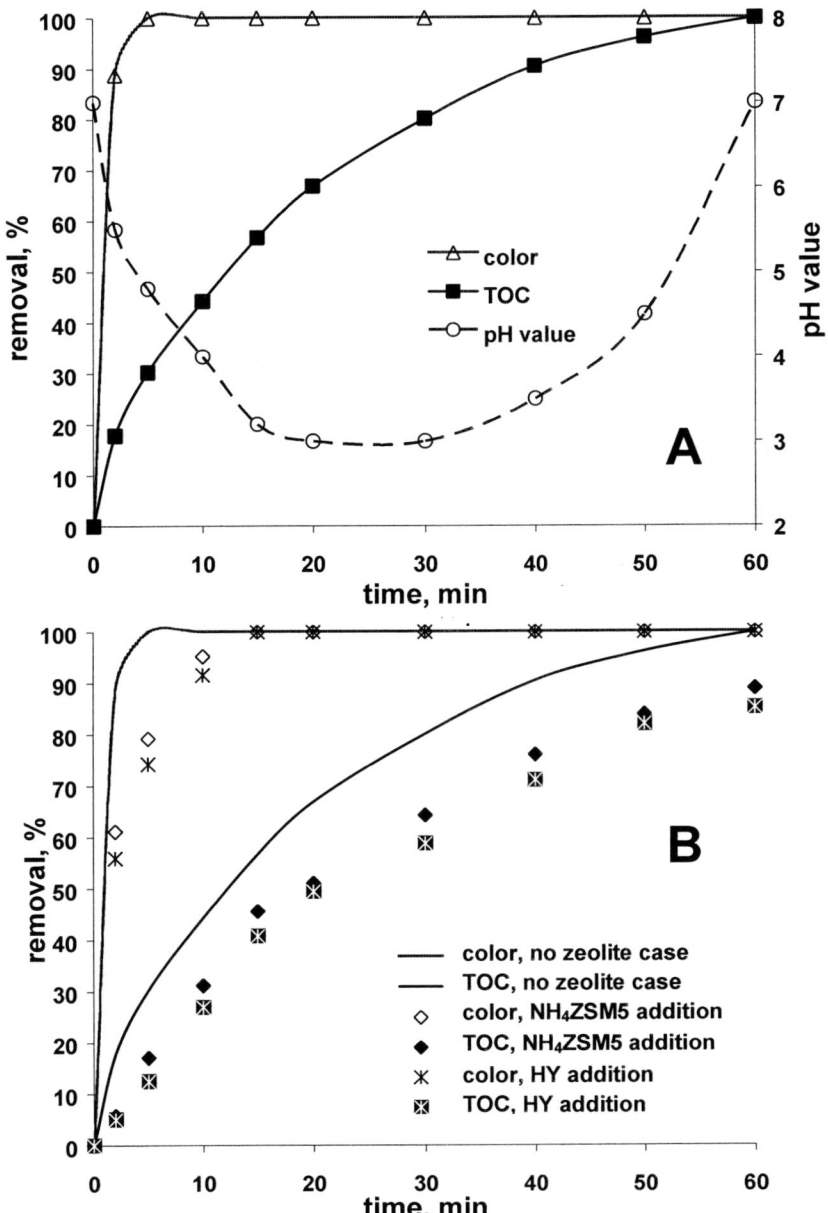

Figure 15. RB137 (20 mgL^{-1}) decolorization and mineralization efficiency as well as changes of pH values during the treatment (A), and influence of NH$_4$ZSM5 and HY zeolites on RB137 degradation efficiency (B) by UV/O$_3$/H$_2$O$_2$ process (pH 7 and c(H$_2$O$_2$)=1.0 mM).

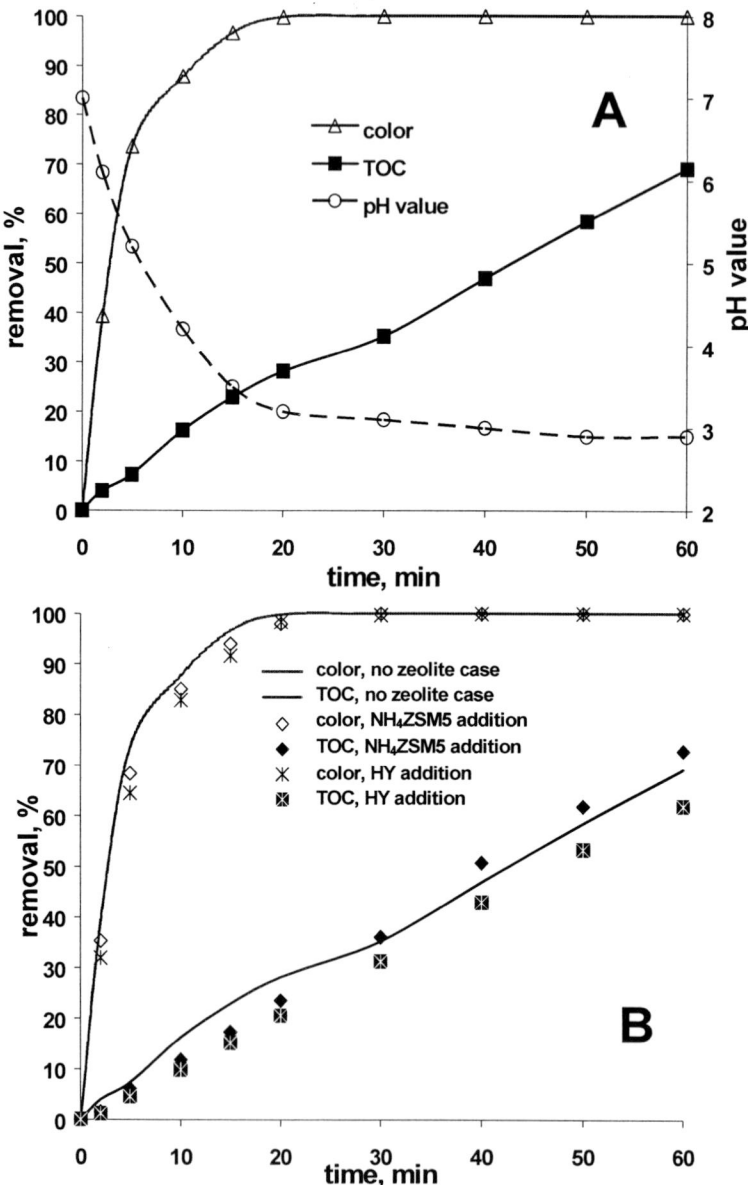

Figure 16. RB137 (80 mgL^{-1}) decolorization and mineralization efficiency as well as changes of pH values during the treatment (A), and influence of NH$_4$ZSM5 and HY zeolites on RB137 degradation efficiency (B) by UV/O$_3$/H$_2$O$_2$ process (pH 7 and c(H$_2$O$_2$)=1.0 mM).

Complete bleaching with non-zeolite case was achieved after 20 minutes of treatment, trend of the curve which connects points of measured TOC removals is almost linear, zeolite addition caused delay in decolorization of RB137, while NH$_4$ZSM5 zeolite addition showed positive effect to TOC removal after 30. minute of process. But it should be pointed out that final achieved mineralization, 69.3 % TOC removal, by UV/O$_3$/H$_2$O$_2$ process is somewhat higher than that obtained by UV/O$_3$ process, 64.4 % TOC removal.

AOPs Cost Estimation

The efficiency of studied AOPs in laboratory scale: Fe^0/H$_2$O$_2$, UV/Fe^0/H$_2$O$_2$, UV/O$_3$ and UV/O$_3$/H$_2$O$_2$, for the treatment of RB137 model wastewater was estimated from the point of view of required operating costs. The evaluation was performed on the basis of the obtained results in the case of higher dye loading in colored model wastewater. It should be noted that evaluation of only operating costs was performed. The overall costs are represented by the sum of the capital costs, the operating costs and the maintenance. For a full-scale system these costs will, besides the nature and the concentration of pollutants, strongly depend on the flow rate of the effluent and the configuration of the reactor [38]. In order to evaluate the operating costs for proposed processes, the costs of used reagents valid for Croatian market were taken into account (table 5).

Table 5. Cost of reagents used in applied processes

reagent	basis	cost (US$)
Fe^0	100 g	13.62
H$_2$O$_2$, 30 %	1 L	37.59
O$_2$, >99.5 %	100 L	0.23
electricity	1 kWh	0.16

Note: costs are given in US$ and the values are valid for Croatian market.

The main estimation results are values of ratio *mineralization extent/process operating costs* (ME/POC). Besides costs estimation on the basis of mineralized organic content, the cost for decolorization below visibility limit, <1 mgL^{-1} that corresponds to decolorization below 1.25 % in our case, are also showed. From the point of view of the most suitable processes for decolorization of RB137 model wastewater below visibility limit, it can be seen that processes combining ozone and UV irradiation, UV/O$_3$ and UV/O$_3$/H$_2$O$_2$, are double cheaper than processes using iron powder, and more important, decolorization is even faster. Moreover, by Fe^0/H_2O_2 process the complete decolorization was not achieved even after a one-hour treatment.

From the ratio ME/POC point of view, it can be seen that processes using UV irradiation are much more efficient, with ME/POC ratios ranged between 3.20-3.45 US\$ g^{-1}, than that one conducted in dark. Moreover, UV irradiation could provide complete mineralization, which is not the case when only degradation over OH radicals is used. However, processes combining UV and ozone seem to be more appropriate due to that successful maintenance of $UV/Fe^0/H_2O_2$ process required strong acid conditions, pH 3. That could increase operating costs of treatment process, but also includes post-treatment as necessary due to the strong conditioning to the neutral pH, which also causes additional costs. Furthermore, post-treatment after the treatment by $UV/Fe^0/H_2O_2$ process includes removal of non-reacted iron powder and remaining iron ions from the bulk. On the contrary, UV/O_3 and $UV/O_3/H_2O_2$ processes were maintained at neutral or weak basic conditions, and after the complete mineralization, the pH values will be the same as at the beginning of treatment process. That excludes additional costs for the strong conditioning of solution before and after the treatment. The cost of processes using UV irradiation could be drastically reduced if solar light is used instead artificial UV.

Table 6. Cost evaluation for all applied processes
(processes conducted in laboratory scale)

process	time required for decolorization <1mgL^{-1} (min)	operational cost for decolorization <1mgL^{-1} (US\$)	final TOC removal (%)	costs (US\$ g^{-1})
Fe^0/H_2O_2	>60	0.046*	19.0	9.37
$UV/Fe^0/H_2O_2$	<30	0.055	80.3	3.20
UV/O_3	<20	0.025	64.4	3.45
$UV/H_2O_2/O_3$	<20	0.028	69.3	3.31

* indicates treatment time over 60 minutes.

CONCLUSION

This chapter is dedicated to the investigation of highly efficient and destructive methods for the minimization of hazardous liquid waste containing reactive dyes as parent molecules and other contaminants present in such wastewaters. Besides an overview of recent publications concerning worldwide literature about AOPs for colored wastewaters treatment given in this chapter, a large part of our scientific investigations up to now, related to this subject, is presented too. The chapter describes different AOPs and the possibility of their usage for the treatment of colored wastewater containing reactive dyes in order to achieve colorless and non-toxic wastewater effluents at first, and, if it is possible, to reach significant minimization of overall organic content. In the research part of this chapter, several AOPs; Fe^0/H_2O_2, $UV/Fe^0/H_2O_2$, UV/O_3 and $UV/O_3/H_2O_2$, are investigated on degradation of C.I. Reactive Blue 137, azo reactive dye, as model pollutant in colored wastewater. It was shown that studied AOPs are very efficient for the bleaching of RB137 model solution. That is also reported through the review part of this chapter, where chemical and photochemical processes shown as the most efficient among all AOPs. Those findings are very important due to the fact that the present color in wastewaters could cause serious problems to aquatic life in recipient waters. Moreover, after decolorization less complex structures are formed that can be degraded by microorganisms, which is not the case when reactive dyes are present in water as parent molecules. Although AOPs seem to be much more expensive to start than common commercial treatment methods such as filtration, adsorption and biodegradation, they are highly efficient for the degradation of recalcitrant pollutants, such as reactive dyes. Moreover, evaluated values of ratio *mineralization extent/process operating costs* (ME/POC) for $UV/Fe^0/H_2O_2$, UV/O_3 and $UV/O_3/H_2O_2$ processes in the case of RB137 model solution treatment ranged between 3.20-3.45 US$ per gram of mineralized organic content. That seems reasonable if complete mineralization considers, i.e. complete removal of toxic organic substances from wastewater by destructive mechanism of AOPs.

ACKNOWLEDGEMENT

We would like to acknowledge financial support from the Ministry of Science, Education and Sport, Republic of Croatia, Project # 0125-018. We are gratefully acknowledged for the support from the National Science Foundation (USA) (INT-0086351) for founding the ozone generator and synthetic zeolites.

REFERENCES

[1] Zheng, Z; Levin, RE; Pinkham, JL; Shetty, K. Decolorization of polymeric dyes by a novel penicillium isolate. *Process Biochemistry,* 1999 34, 31-37.

[2] Zollinger, H. *Colour Chemistry – Synthesis, Properties and Applications of Organic Dyes and Pigments.* New York: VHC Publishers; 1987.

[3] Riefe, A; Freeman, HS. *Environmental Chemistry of Dyes and Pigments.* New York: John Wiley and Sons; 1996.

[4] Phillips, D. Environmentally friendly, productive and reliable: priorities for cotton dyes and processes. *Journal of Society of Dyers and Colourists,* 1996 112, 183-186.

[5] Hunger, K. *Industrial Dyes; Chemistry, Properties, Application.* Weinheim: Wiley-VCH; 2002.

[6] Allen, SJ; Khader, KYH; Bino, M. Electrooxidation of dyestuff in waste waters. *Journal of Chemical and Biochemical Technology,* 1999 62, 111-117.

[7] Wang, C; Yediler, A; Lienert, D; Wang, Z; Kettrup, A. Toxicity evaluation of reactive dyestuffs, auxiliaries and selected effluents in textile finishing industry to luminescent bacteria Vibrio fischeri. *Chemosphere,* 2002 46 (2), 339-344.

[8] Gottlieb, A; Shaw, C; Smith, A; Wheatley, A; Forsythe, S. The toxicity of textile reactive azo dyes after hydrolysis and decolourisation. *Journal of Biotechnology,* 2003 101(1), 49-56.

[9] Nillson, R; Nordlinder, R; Wass, U. Asthma, rhinitis, and dermatitis in workers exposed to reactive dyes. *British Journal of Industrial Medicine,* 1993 50, 65-70.

[10] Arpe, H-J. Ullmann's Encyclopedia of Industrial Chemistry: Volume A2, Amines, Aliphatic to Antibiotics. 5th Edition, Weinheim: Wiley-VCH; 1985.
[11] Kelshaw, P. The future for West European dyestuff manufacturers. *Journal of Society of Dyers and Colourists,* 1998 114, 35-37.
[12] Sincero, AP; Sincero, GA. *Physical-Chemical Treatment of Water and Wastewater.* New York: CRC Press, IWA Publishing; 2003.
[13] Droste, RJ. *Theory and Practice of Water and Wastewater Treatment.* New York: John Wiley and Sons; 1997.
[14] Forgacs, E; Cserhati, T; Oros, G. Removal of synthetic dyes from wastewaters: a review. *Environmental International,* 2004 30, 953-971.
[15] Pagga, U; Brown, D. The degradation of dyestuffs: part ii behavior of dyestuffs in aerobic biodegradation tests. *Chemosphere,* 1986 15, 479–491.
[16] Seshadri, S; Bishop, PI; Agha, AM. Anaerobic/aerobic treatment of selected azo dyes in wastewater. *Waste Management,* 1994 15, 127–137.
[17] Lourenço, ND; Novais, JM; Pinheiro, HM. Effect of some operational parameters on textile dye biodegradation in a sequential batch reactor. *Journal of Biotechnology,* 2001 89, 163-174.
[18] Georgiou, D; Hatiras, J; Aivasidis, A. Microbial immobilization in a two-stage fixed-bed-reactor pilot plant for on-site anaerobic decolorization of textile wastewater. *Enzyme and Microbial Technology,* 2005 37 (6) 597-605.
[19] Ambrósio, ST; Campos-Takaki, GM. Decolorization of reactive azo dyes by Cunninghamella elegans UCP 542 under co-metabolic conditions. *Bioresource Technology,* 2004 91, 69-75.
[20] Ozdemir, O; Armagan, B. Turan, M; Celik, MS. Comparison of the adsorption characteristics of azo-reactive dyes on mezoporous minerals. *Dyes and Pigments,* 2004 62, 49-60.
[21] Karcher, S; Kornmüller, A; Jekel, M. Screening of commercial sorbents for the removal of reactive dyes. *Dyes and Pigments,* 2001 51, 111-125.
[22] Wang, S; Li, H; Xu, L. Application of zeolite MCM-22 for basic dye removal from wastewater. *Journal of Colloid and Interface Science,* 2006 295, 71-78.
[23] Benkli, YE; Can, MF; Turan, M; Çelik, MS. Modification of organo-zeolite surface for the removal of reactive azo dyes in fixed-bed reactors. *Water Research,* 2005 39, 487-493.
[24] Armaan, B; Turan, M; Cęlik, MS. Equilibrium studies on the adsorption of reactive azodyes into zeolite. *Desalination,* 2004 170, 33-39.

[25] Papić, S; Koprivanac, N; Lončarić Božić, A. Removal of reactive dyes from wastewater using Fe(III) coagulant. *Journal of Society of Dyers and Colourist,* 2000 116, 352-358.
[26] Koprivanac, N, Lončarić Božić, A, Papić, S. Cleaner production processes in the synthesis of blue anthraquinone reactive dye. *Dyes and Pigments,* 2000 44, 33-40.
[27] Papić, S; Koprivanac, N; Lončarić Božić, A; Meteš A. Removal of some reactive dyes from synthetic wastewater by combined Al(III) coagulation/carbon adsorption process. *Dyes and Pigments,* 2004 62, 293-300.
[28] White, GC. Handbook of Chlorination and Alternative Disinfectants. New York: John Wiley & Sons; 1999.
[29] La Grega, MD; Buckingham, PL; Evans, JC. *Hazardous Waste Management.* New York: McGraw-Hill; 1994.
[30] Chu, W; Ma, C-W. Quantitative prediction of direct and indirect dye ozonation kinetics. *Water Research,* 2000 34 (12), 3153-3160.
[31] Parsons, SA; Williams, M. Introduction. In: Parsons SA, editor. *Advanced oxidation processes for water and wastewater treatment.* London: IWA Publishing; 2004; 1-6.
[32] Gogate, PR; Pandit, AB. A review of imperative technologies for wastewater treatment I: oxidation technologies at ambient conditions. *Advances in Environmental Research,* 2004 8, 501-551.
[33] Gogate, PR; Pandit, AB. A review of imperative technologies for wastewater treatment II: hybrid methods. *Advances in Environmental Research,* 2004 8, 553-597.
[34] Glaze, WH; Kang, JW; Chapin, DH. The chemistry of water treatment processes involving ozone, hydrogen peroxide and ultraviolet irradiation. *Ozone Science and Engineering,* 1987 9, 335–352.
[35] Beltran, FJ. Ozone-UV radiation-hydrogen peroxide oxidation technologies. In: Tarr MA, editor. *Chemical degradation methods for wastes and pollutants - environmental and industrial applications.* New York: Marcel Dekker; 2003; 1-75.
[36] Rodriguez, M. Fenton and UV-VIS based advanced oxidation processes in wastewater treatment: Degradation, mineralization and biodegradability enhancement. PhD Thesis. Barcelona: University of Barcelona, 2003.
[37] Bigda, RJ. Consider Fenton's chemistry for wastewater treatment. *Chemical Engineering Progressing,* 1995 91, 62-66.

[38] Andreozzi, R; Vaprio, V; Insola, A; Marotta, R. Advanced oxidation processes (AOP) for water purification and recovery. *Catalysis Today*, 1999 53, 51-59.
[39] Grymonpre, DR; Finney, WC; Clark, RJ; Locke BR. Hybrid gas-liquid electrical discharge reactors for organic compound degradation, *Industrial and Engineering Chemistry Research*, 2004 43, 1975-1989.
[40] Locke, BR; Sato, M; Sunka, P; Hoffmann, MR, Chang; JS. Electrohydraulic discharge and nonthermal plasma for water treatment. *Industrial and Engineering Chemistry Research*, 2006 45 (3), 882-905.
[41] Fenton, HJH. On a new reaction of tartaric acid. *Chemical News*, 1876 33, 190.
[42] Fenton, HJH. Oxidation of tartaric acid in presence of iron. *Journal of Chemical Society*, 1894 65, 899-910.
[43] Tarr, MA. Fenton and modified Fenton methods for pollutant degradation. In: Tarr MA, editor. *Chemical degradation methods for wastes and pollutants - environmental and industrial applications*. New York: Marcel Dekker; 2003; 1-77.
[44] Xu, X-R; Li, H-B; Wang, W-H; Gu, J-D. Degradation of dyes in aqueous solutions by the Fenton process. *Chemosphere*, 2004 57, 595-600.
[45] Lucas, MS; Peres JA. Decolorization of the azo dye Reactive Black 5 by Fenton and photo-Fenton oxidation. *Dyes and Pigments*, 2006 71, 236-244.
[46] Meric, S; Selcuk, H; Gallo, M; Belgiorno V. Decolourisation and detoxifying of Remazol Red dye and its mixture using Fenton's reagent. *Desalination*, 2005 173, 239-248
[47] Chen, R; Pignatello, JJ. Role of quinone intermediates as electron shuttles in Fenton and photoassisted Fenton oxidations of aromatic compounds. *Environmental Science Technology*, 1997 31, 2399-2406.
[48] Kušić, H; Koprivanac, N; Lončarić Božić, A; Selanec, I. Photo-assisted Fenton type processes for the degradation of phenol: a kinetic study. *Journal of Hazardous Materials*, 2006 in press.
[49] Kusic, H; Loncaric Bozic, A; Koprivanac, N. Fenton type processes for minimization of organic content in coloured wastewaters: Part I: processes optimization. *Dyes and Pigments*, 2006 in press.
[50] Gallard, H; De Laat, J. Kinetics of oxidation of chlorobenzenes and phenyl-ureas by $Fe(II)/H_2O_2$ and $Fe(III)/H_2O_2$. Evidence of reduction and oxidation reactions of intermediates by Fe(II) and Fe(III). *Chemosphere*, 2001 42, 405-413.
[51] Neamtu, M; Yediler, A; Siminiceanu, I; Kettrup, A. Oxidation of commercial reactive azo dye aqueous solutions by the photo-Fenton and

Fenton-like processes. *Journal of Photochemistry and Photobiology A: Chemistry,* 2003 161, 87–93.

[52] Kusic, H, Koprivanac, N; Srsan, L. Azo dye degradation using Fenton type processes assisted by UV irradiation: A kinetic study. *Journal of Photochemistry and Photobiology A: Chemistry,* 2006 181 (2-3), 195-202.

[53] Bergendahl, JA; Thies, TP. Fenton's oxidation of MTBE with zero-valent iron. *Water Research,* 2004 38, 327-334.

[54] Tang, WZ; Chen, RZ. Decolorization kinetics and mechanism of commercial dyes by H_2O_2/iron powder system. *Chemosphere,* 1996 32 (5), 947-958.

[55] Costa, RCC; Lelis, MFF; Oliveira, LCA; Fabris, JD; Ardisson, JD; Rios, RRVA; Silva, CN; Lago, RM. Novel active heterogeneous Fenton system based on $Fe_{3-x}M_xO_4$ (Fe, Co, Mn, Ni): The role of M^{2+} species on the reactivity towards H_2O_2 reactions. *Journal of Hazardous Materials,* 2006 129, 171-178.

[56] Moura, FCC; Araujo, MH; Costa, RCC; Fabris, JD; Ardisson, JD; Macedo, WAA; Lago, RM. Efficient use of Fe metal as an electron transfer agent in a heterogeneous Fenton system based on Fe^0/Fe_3O_4 composites. *Chemosphere,* 2005 60, 1118-1123.

[57] He, J; Ma, W; Song, W; Zhao, J; Qian, X; Zhang, S; Yu JC. Photoreaction of aromatic compounds at α-FeOOH/H_2O interface in the presence of H_2O_2: evidence for organic-goethite surface complex formation. *Water Research,* 2005 39, 119-128.

[58] Phu, NH; Hoa, TTK; Tan, NV; Thang HV; Ha, PL. Characterization and activity of Fe-ZSM-5 catalysts for the total oxidation of phenol in aqueous solutions. *Applied Catalysis, B: Environmental, 2001* 34, 267-275.

[59] Kušić, H; Koprivanac, N; Selanec, I. Fe-exchanged zeolite as the effective heterogeneous Fenton-type catalyst for the organic pollutant minimization: UV irradiation assistance, *Chemosphere,* 2006 *in press*

[60] Catrinescu, C; Neamtu, M; Yediler, A; Macoveanu, M; Kettrup, A. Catalytic wet peroxide oxidation of an azo dye, Reactive Yellow 84, over Fe-exchanged ultrastable Y zeolite. *Environmental Engineering and Management Journal,* 2002 1, 177-186.

[61] Catrinescu, C; Teodosiu, C, Macoveanu, M; Miehe-Brendlé, J; Le Dred, R. Catalytic wet peroxide oxidation of phenol over Fe-exchanged pillared beidellite. *Water Research,* 2003 37, 1154-1160.

[62] Ly, X; Xu, Y; Ly, K; Zhang, G. Photo-assisted degradation of anionic and cationic dyes over iron(III)-loaded resin in the presence of hydrogen

peroxide. *Journal of Photochemistry and Photobiology, A: Chemistry,* 2005 173, 121-127.

[63] Verma, P; Baldrian, P; Nerud, F. Decolorization of structurally different synthetic dyes using cobalt(II)/ascorbic acid/hydrogen peroxide system. *Chemosphere,* 2003 50, 975-979

[64] Kavitha, V; Palanivelu, K. The role of ferrous ion in Fenton and photo-Fenton processes for the degradation of phenol. *Chemosphere,* 2004 55, 1235-1243.

[65] www.h2o2.com

[66] Arslan, I; Balcioglu, IA. Degradation of Remazol Black B dye and its simulated dyebath wastewater by advanced oxidation processes in heterogeneous and homogeneous media. *Coloration Technology,* 2001 117, 38-42.

[67] Hsueh, CL; Huang, YH; Wang, CC; Chen; CY. Degradation of azo dyes using low iron concentration of Fenton and Fenton-like system. *Chemosphere,* 2005 58 (10), 1409-1414.

[68] Carneiro, PA; Pupo Nogueira, RF; Zanoni, MVB. Homogeneous photodegradation of C.I. Reactive Blue 4 using a photo-Fenton process under artificial and solar irradiation. *Dyes and Pigments,* 2006 *in press.*

[69] Hsueh, C-L; Huanga, Y-H; Wang, C-C; Chena, C-Y. Photoassisted Fenton degradation of nonbiodegradable azo-dye (Reactive Black 5) over a novel supported iron oxide catalyst at neutral pH. *Journal of Molecular Catalysis A: Chemical,* 2006 245, 78–86.

[70] Muruganandham, M; Swaminathan, M. Decolourisation of Reactive Orange 4 by Fenton and photo-Fenton oxidation technology. Dyes and Pigments, 2004 63, 315-321.

[71] Stylidi, M; Kondarides, DI; Verykios, XE. Pathways of solar light-induced photocatalytic degradation of azo dyes in aqueous TiO_2 suspensions. *Applied Catalysis B: Environmental,* 2003 40, 271-286.

[72] Stefan, MI. UV photolysis: background. In: Parsons SA, editor. *Advanced oxidation processes for water and wastewater treatment.* London: IWA Publishing; 2004; 7-48.

[73] Legrini, O; Oliveros, E; Braun, AM. Photochemical process for water treatment. *Chemical Review,* 1993 93 (2), 671-698.

[74] Arslan, I; Balcioglu, IA; Bahnemann, D. Advanced chemical oxidation of reactive dyes in simulated dyehouse effluents by ferrioxalate–Fenton/UV-A and TiO_2/UV-A processes. *Water Research,* 2001 37, 3061-3069.

[75] Muruganandham, M; Swaminathan, M. Solar driven decolourisation of Reactive Yellow 14 by advanced oxidation processes in heterogeneous and homogeneous media. *Dyes and Pigments,* 2007 72 (2), 137-143.

[76] Bali, U; Catalkaya, E; Sengul F. Photodegradation of Reactive Black 5, Direct Red 28 and Direct Yellow 12 using UV, UV/H_2O_2 and UV/H_2O_2/Fe^{2+}: a comparative study. *Journal of Hazardous Materials,* 2004 B114, 159–166.

[77] Behnajady, MA; Modirshahla, N; Fathi H. Kinetics of decolorization of an azo dye in UV alone and UV/H_2O_2 processes. *Journal of Hazardous Materials,* 2006 133 (1-3), 226-232.

[78] Peternel, I; Koprivanac, N; Kusic, H. UV-based processes for reactive azo dye mineralization. *Water Research,* 2006 40 (3), 525-532.

[79] Kušić, H; Koprivanac, N; Lončarić Božić, A; Papić, S; Peternel, I; Vujević, D. Reactive dye degradation by a photochemical AOP – development of a kinetic model for UV/H_2O_2 process. *Chemical and Biochemical Engineering Quarterly,* 2006 in press.

[80] Neppolian, B; Choi, HC; Sakthivel, S; Arabindoo, B; Murugesan, V. Solar/UV-induced photocatalytic degradation of three commercial textile dyes. *Journal of Hazardous Materials,* 2002 B89, 303-317.

[81] Tuhkanen, T. UV/H_2O_2 process. In: Parsons SA, editor. *Advanced oxidation processes for water and wastewater treatment.* London: IWA Publishing; 2004; 86-110.

[82] Hunt, JP; Taube, H. The photochemical decomposition of hydrogen peroxide, Quantum yields, transfer and fractionation effects. *Journal of American Chemical Society,* 1952 74, 5999-6002.

[83] Eckenfelder, WW. *Industrial Water Pollution Control.* 2nd Edition. New York: McGraw-Hill; 1989.

[84] Crittenden, C; Hu, S; Hand, DW; Green, SA. A kinetic model for H_2O_2/UV process in a completely mixed batch reactor. *Water Research,* 1999 33 (10), 2315-2328.

[85] De, AK; Chaudhuri, B; Bhattacharjee, S; Dutta, BK. Estimation of OH$^\bullet$ radical reaction rate constants for phenol and chlorinated phenols using UV/H_2O_2 photo-oxidation. *Journal of Hazardous Materials,* 1999 64 (1), 91-104.

[86] Guittonneau, S; De Laat, J; Duguet, JP; Bonnel, C; Dore, M. Oxidation of parachloronitrobenzene in dilute Aqueous solution by O_3 + UV and H_2O_2 + UV: A comparative study, *Ozone Science and Engineering,* 1990 12, 73-94.

[87] Kusic, H; Koprivanac, N; Loncaric Bozic, A. Minimization of organic pollutant content in aqueous solution by means of AOPs: UV- and ozone–based technologies. *Chemical Engineering Journal*, 2006 in press

[88] Mohey El-Dein, A; Libra, JA; Wiesmann U. Mechanism and kinetic model for the decolorization of the azo dye Reactive Black 5 by hydrogen peroxide and UV radiation. *Chemosphere*, 2003 52, 1069–1077.

[89] Muruganandham, M; Swaminathan M. Photochemical oxidation of reactive azo dye with UV–H_2O_2 process. *Dyes and Pigments*, 2004 62, 269–275.

[90] Colonna, GM; Caronna, T; Marcandalli, B. Oxidative degradation of dyes by ultraviolet radiation in the presence of hydrogen peroxide. *Dyes and Pigments*, 1999 41, 211-220.

[91] Ince, N. Critical effect of hydrogen peroxide in photochemical dye degradation. *Water Research*, 1999 33, 1080-1084.

[92] Shu, H; Chang, M. Decolorization effects of six azo dyes by O_3, UV/O_3 and UV/H_2O_2 processes. *Dyes and Pigments*, 2005 65, 25-31.

[93] Wu, C-H; Chang, C-L. Decolorization of Reactive Red 2 by advanced oxidation processes: Comparative studies of homogeneous and heterogeneous systems. *Journal of Hazardous Materials*, 2006 B128, 265–272.

[94] Mills, A; Lee, S-K. Semiconductor photocatalysis. In: Parsons SA, editor. *Advanced oxidation processes for water and wastewater treatment.* London: IWA Publishing; 2004; 137-166.

[95] Pichat, P. Photocatalytic degradation of pollutants in water and air:basic concepts and applications. In: Tarr MA, editor. *Chemical degradation methods for wastes and pollutants - environmental and industrial applications.* New York: Marcel Dekker; 2003; 77-119.

[96] Carneiro, PA; Osugi ME; Sene, JJ; Anderson, MA; Boldrin Zanoni, MV. Evaluation of color removal and degradation of a reactive textile azo dye on nanoporous TiO_2 thin-film electrodes. *Electrochimica Acta*, 2004 49, 3807–3820.

[97] Goncalves, MST; Pinto, EMS; Nkeonye, P; Oliveira-Campos, AMF. Degradation of C.I. Reactive Orange 4 and its simulated dyebath wastewater by heterogeneous photocatalysis. *Dyes and Pigments*, 2005 64, 135-139.

[98] Reutergradh, LB; Iangpashuk, M. Photocatalytic decolorization of reactive azo dye: a Comparison between TiO_2 and CdS photocatalysis, *Chemosphere*, 1997 35, 585-596.

[99] Daneshvar, N; Salari, D; Khataee, AR. Photocatalytic degradation of azo dye Acid Red 14 in water on ZnO as an alternative catalyst to TiO_2.

Journal Photochemistry and Photobiology A: Chemistry, 2004 162, 317-322.

[100] Sakthivel, S; Neppolian, B; Shankar, MV; Arabindoo, B; Palanichamy, M, Murugesan, V. Solar photocatalytic degradation of azo dye: Comparison of photocatalytic efficiency of ZnO and TiO_2. *Solar Energy Materials and Solar Cells,* 2003 77, 65-82.

[101] Vinogdopal, K; Wynkoop, DE; Kamat, PV. Environmental chemistry on semiconductor surfaces: Photosensitized degradation of a textile azo dye, Acid Orange 7, on the TiO_2 particles using visible light. *Environmental Science and Technology,* 1996 30, 1660-1666.

[102] Bizani, E; Fytianos, K; Poulios I; Tsiridis V. Photocatalytic decolorization and degradation of dye solutions and wastewaters in the presence of titanium dioxide. *Journal of Hazardous Materials,* 2006 *in press.*

[103] Muruganandham, M; Swaminathan M. TiO_2–UV photocatalytic oxidation of Reactive Yellow 14: Effect of operational parameters. *Journal of Hazardous Materials,* 2006 *in press.*

[104] Peternel, I; Koprivanac, N; Loncaric Bozic, A; Kusic, H. Comparative study of UV/TiO_2, UV/ZnO and UV/Fenton processes for the organic reactive dye in aqueous solution. *Waste Management,* 2006 *submitted*

[105] Akyol A; Yatmaz, HC; Bayramoglu, M. Photocatalytic decolorization of Remazol Red RR in aqueous ZnO suspensions. *Applied Catalysis B: Environmental,* 2004 54, 19–24.

[106] www.medicina.hr

[107] www.tifm.com/uvbozone

[108] Langlais, B; Reckhow, DA; Brink, DR. *Ozone in water treatment; application and engineering.* Boca Raton: Lewis Publishers; 1991.

[109] Tomiyasu, H; Fukutomi, H; Gordon, G. Kinetics and mechanism of ozone decomposition in basic aqueous solution. *Inorganic Chemistry,* 1985 24 (19), 2962-2966.

[110] Mathews, AP; Panda, KK; Ananthi, S; Padmanabhan, K. Mass transfer and oxidation kinetics in an in situ ozone generator. *Water Science and Technology,* 2004 49, 13-18.

[111] Arslan-Alaton, I; Dogruel, S; Baykal, E; Gerone, G. Combined chemical and biological oxidation of penicillin formulation effluent. *Journal of Environmental Management,* 2004 73 (2), 155-163.

[112] Glaze, W; Kang, J. Advanced oxidation processes. Test of a kinetic model for the oxidation of organic compounds with ozone and hydrogen peroxide in a semi batch reactor, *Industrial and Engineering Chemistry Research,* 1989 28, 1580-1587.

[113] Wu, J; Wang, T. Ozonation of aqueous azo dye in semi-batch reactor. *Water Research,* 2001 35 (4), 1093-1099.
[114] Peralta–Zamora, P; Kunz, A; Gomes de Moreas, S; Pelegrini, R; de Campos Moleiro, P; Reyes, J; Durán, N. Degradation of reactive dyes I. a comparative study of ozonation, enzymatic and photochemical processes. *Chemosphere,* 1999 38 (4), 835-852.
[115] Arslan-Alaton, I; Bacioglu, IA; Bahnemann, DW. Advanced oxidation of a reactive dyebath effluent: comparison of O_3, H_2O_2/UV-C and TiO_2/UV-A processes. *Water Research,* 2002 36, 1143-1154.
[116] Arslan, I; Bacioglu, IA. Effect of common reactive dye auxiliaries on the ozonation of dyehouse effluents contaning vinylsulphone and aminochlorotriazine dyes. *Desalination,* 2000 130, 61-71.
[117] Kušić, H; Lončarić Božić, A; Koprivanac, N; Peternel, I; Vujević, D; Papić, S. Comparative study of several ozone based processes for reactive dye degradation. In: Kruithof J, editor. Proceedings of 17th world congress & exhibition: Ozone & Related Oxidants, Innovative & Current Technologies. *Strasbourg: International Ozone Association;* 2005; VII.3.24.1-10.
[118] Ince, NH; Tezcanli, G. Reactive dyestuff degradation by combined sonolysis and ozonation. *Dyes and Pigments,* 2001 49 (3), 145-153.
[119] Lukes, P. *Water treatment by pulsed streamer corona discharge.* PhD Thesis. Prague: Institute of Plasma Physics AS CR; 2001.
[120] Joshi, AA; Locke, BR; Arce, P; Finney, WC. Formation of hydroxyl radicals, hydrogen peroxide and aqueous electrons by pulsed streamer corona discharge in aqueous solution. *Journal of Hazardous Materials,* 1995 41, 3-30.
[121] Sunka, P; Babicky, V; Clupek, M; Lukes, P; Simek, M; Schmidt, J; Cernak, M. Generation of chemically active species by electrical discharges in water. *Plasma Sources Science and Technology,* 1999 8 (2), 258-265.
[122] Clements, JS; Sato, M; Davis, RH. Preliminary investigation of prebreakdown phenomena and chemical reactions using a pulsed high-voltage discharge in water. IEEE Trans. *Industrial Applications,* 1987 IA-23, 224-235.
[123] Sato, M; Sun, B; Ohshima, T; Sagi, Y. Characteristics of active species and removal of organic compounds by a pulsed corona discharge in water. *Journal of Advanced Oxidation Technologies,* 1999 4, 339-342.

[124] Sun, B; Sato, M; Clements, JS. Optical study of active species produced by a pulsed streamer corona discharge in water. *Journal of Electrostatics,* 1997 39, 189-202.

[125] Grymonpre, DR. *An experimental and theoretical analysis of phenol degradation by pulsed corona discharge.* PhD Thesis. Tallahassee: Florida State University; 2001.

[126] van Craeynest, K; van Langenhove, H; Stuetz, RM. AOPs for VOCs and odour treatment: Non-thermal palsma. In: Parsons SA, editor. *Advanced oxidation processes for water and wastewater treatment.* London: IWA Publishing; 2004; 275-301.

[127] Lukes, P; Appleton, AT; Locke, BR. Hydrogen peroxide and ozone formation in hybrid gas-liquid electrical discharge reactors. *IEEE Trans. Industrial. Applications,* 2004 40 (1), 60–67.

[128] Sugiarto, AT; Ohshima, T; Sato, M. Advanced oxidation processes using pulsed streamer corona discharge in water. *Thin Solid Films,* 2002 407, 174-178.

[129] Sunka, P; Babicky, V; Clupek, M; Fuciman, M; Lukes, P; Simek, M; Benes, J; Locke, BR; Majcherova, Z. Potential applications of pulse electrical discharges in water, *Acta Physica Slovaca,* 2004 54 (2), 135-145.

[130] Loncaric Bozic, A; Koprivanac, N; Sunka, P, Clupek, M; Babicky, V. Organic synthetic dye degradation by modified pinhole discharge. *Czechoslovak Journal of Physics,* 2004 54C, C958-C963.

[131] Koprivanac, N; Kusic, H; Vujevic, D; Peterenel, I; Locke, BR. Influence of iron on degradation of organic dyes in corona. *Journal of Hazardous Materials,* 2005 B117, 113-119.

[132] Kusic, H; Koprivanac, N; Peternel, I; Locke BR. Hybrid gas/liquid electrical discharge reactors with zeolites for colored wastewater degradation. *Journal of Advanced Oxidation Technologies,* 2005 8 (2), 172-181.

[133] Peternel, I; Kusic, H; Koprivanac, N; Locke, BR. The roles of ozone and zeolite on reactive dye degradation in electrical discharge reactors. *Environmental Technology,* 2006 27 (5) 545-558.

[134] Mason, TJ; Petrier, C. Ultrasound processes. In: Parsons SA, editor. *Advanced oxidation processes for water and wastewater treatment.* London: IWA Publishing; 2004; 185-208.

[135] Destaillats, H; Hoffmann, MR; Wallace, HC. Sonochemical degradation of polutants. In: Tarr MA, editor. *Chemical degradation methods for wastes and pollutants - environmental and industrial applications.* New York: Marcel Dekker; 2003; 201-233.

[136] Vajnhandl, S; Majcen Le Marechal, A. Ultrasound in textile dyeing and the decolouration/mineralization of textile dyes. *Dyes and Pigments,* 2005 65, 89-101.
[137] Tezcanli-Guyer, G. Ince, NH. Degradation and toxicity reduction of textile dyestuff by ultrasound. *Ultrasonics Sonochemistry,* 2003 10, 235–240.
[138] Rehorek, A; Tauber M; Gubitz G. Application of power ultrasound for azo dye degradation. *Ultrasonics Sonochemistry,* 2004 11, 177–182.
[139] Cooper, WJ; Gehringer, P; Pikaev, AK; Kurcuz, CN; Mincher, BJ. Radiation processes. In: Parsons SA, editor. *Advanced oxidation processes for water and wastewater treatment.* London: IWA Publishing; 2004; 209-248.
[140] Dajka, K; Takacs, E; Solpan, D; Wojnarovits, L; Guven, O. High-energy irradiation treatment of aqueous solutions of C.I. Reactive Black 5 azo dye: pulse radiolysis experiments. *Radiation Physics and Chemistry,* 2003 67, 535–538.
[141] Zhang, S-J; Yu, H-Q; Li, Q-R. Radiolytic degradation of Acid Orange 7: A mechanistic study. *Chemosphere,* 2005 61, 1003–1011.
[142] Brillas, E; Cabot, P-L; Casado, J. Electrchemical methods for degradation of organic pollutants in aqueous media. In: Tarr MA, editor. *Chemical degradation methods for wastes and pollutants - environmental and industrial applications.* New York: Marcel Dekker; 2003; 235-304.
[143] Catanho, M; Malpass, GRP; Motheo AJ. Photoelectrochemical treatment of the dye reactive red 198 using DSA electrodes. *Applied Catalysis B: Environmental,* 2006 62, 193–200.
[144] Awad, HS; Abo Galwa N. Electrochemical degradation of Acid Blue and Basic Brown dyes on Pb/PbO_2 electrode in the presence of different conductive electrolyte and effect of various operating factors. *Chemosphere,* 2005 61, 1327–1335.
[145] Carneiro, PA; Osugi, ME; Fugivara, CS; Boralle, N; Furlan, M; Zanoni, MVB. Evaluation of different electrochemical methods on the oxidation and degradation of Reactive Blue 4 in aqueous solution. *Chemosphere,* 2005 59, 431–439.
[146] Kusic, H; Lonacaric Bozic, A; Koprivanac, N; Papic, S. Fenton type processes for minimization of organic content in coloured wastewaters: Part II: combination with zeolites. Dyes and Pigments, 2006 *in press*
[147] Papić, S; Koprivanac, N; Lončarić Božić, A; Vujević, D; Dragičević Kučar, S; Kušić, H; Peternel, I. AOPs in azo dye wastewater treatment. Water Environment Research, 2006 78 (6), 572-579.

[148] Tomlinson, AAG. *Modern zeolites structure and function in detergents and petrochemicals*. Zurich; Trans Tech Publications; 1998.
[149] www.bza.org/zeolites
[150] www.zeolyst.com
[151] Nicole, I; De Laat, J; Dore, M, Duguet, JP; Bonnel, C. Utilisation du rayonnement ultraviolet dans le traitement des eaux: mesure du flux photonique par actinometrie chimique au peroxyde d'hydrogene: Use of U.V. radiation in water treatment: measurement of photonic flux by hydrogen peroxide actinometry. *Water Research,* 1990 24 (2), 157-168.
[152] APHA Standard methods for the examination of water and wastewater treatment. 20th Edition. Washington DC: American Public Health Association; 1998.
[153] Pretsch, E; Seibl, J; Simon, W. *Tabellen zur strukturaufklarung organischer verbindungen mit spektroskopischen methoden.* Berlin: Springer Verlag; 1981.
[154] Alaton, IA; Balcioglu, IA; Bahnemann, D. Advanced oxidation of reactive dyebath effluent: comparison of O_3, H_2O_2/UV-A and TiO_2/UV-A processes. *Water Research,* 2002 36, 1143-1154.
[155] Gutowska, A; Kaluzna-Czaplinska, J; Jozwiak WK. Degradation mechanism of Reactive Orange 113 dye by H_2O_2/Fe^{2+} and ozone in aqueous solution. *Dyes and Pigments,* 2006 *in press*.
[156] Feng, W; Nansheng, D; Helin H. Degradation mechanism of azo dye C. I. reactive red 2 by iron powder reduction and photooxidation in aqueous solutions. *Chemosphere*, 2000 41, 1233-1238.
[157] Kawai, T; Tsutsumi, K. Adsorption characteristic of surfactants and phenol on modified zeolites from their aqueous solutions. *Colloid Polymeric Science,* 1995 273, 787-792.
[158] Kušić, H; Koprivanac, N; Locke, BR. Decomposition of phenol by hybrid gas/liquid electrical discharge reactors with zeolite catalysts. *Journal of Hazardous Materials,* 2005 125, 190-200.
[159] Daneshvar, N; Salari, D; Khataee AR. Photocatalytic degradation of azo dye acid red 14 in water: investigation of the effect of operational parameters. *Journal of Photochemistry and Photobiology A: Chemistry,* 2003 157, 111-116.
[160] Bilgi, S; Demir, C. Identification of photooxidation degradation products of C.I. Reactive Orange 16 dye by gas chromatography - mass spectrometry. *Dyes and Pigments,* 2005 66, 69-76.

[161] Hoigné, J; Bader, H. Rate constants of reactions of ozone with organic and inorganic compounds in water—II: Dissociating organic compounds. *Water Research,* 1983 17 (2), 185-194.

INDEX

A

absorption, 16, 36
absorption coefficient, 16, 36
acceptance, 3
acetaldehyde, 5
acetic acid, 13, 46
acetone, 4
achievement, 31, 33, 50
acid, 4, 13, 32, 43, 45, 46, 50, 54, 62, 64, 71
acidic, 21, 23, 45, 46
acoustic, 28
acoustical, 28
activated carbon, 3, 20
adiabatic, 28
adsorption, 3, 23, 32, 38, 39, 45, 46, 55, 60, 61
aerobic, 3, 60
agent, 3, 21, 33, 63
air, 6, 66
alternative, 4, 19, 66
aluminum, 31
amines, 2, 5
amino, 2
anaerobic, 60
aniline, 5
application, vii, 1, 3, 4, 5, 6, 9, 10, 11, 14, 15, 17, 19, 20, 22, 24, 25, 26, 67
aquatic, 55
aqueous solution, 4, 29, 32, 39, 45, 62, 63, 66, 67, 68, 70, 71
aqueous solutions, 4, 29, 62, 63, 70, 71
aromatic, 2, 17, 23, 36, 40, 42, 45, 62, 63
aromatic compounds, 36, 62, 63
aromatic rings, 42
aromatics, 22, 28, 40
artificial, 54, 64
ascorbic, 64
ascorbic acid, 64
atmosphere, 20
atmospheric pressure, 5
atoms, 20, 29
avoidance, 12
azo dye, vii, 2, 3, 11, 32, 39, 59, 60, 62, 63, 64, 65, 66, 67, 68, 70, 71

B

bacteria, 1, 59
Bali, 15, 17, 65
barrier, 26
beams, 29
behavior, 34, 45, 46, 49, 50, 60
benefits, 10, 12
benign, 11
benzene, 5

benzoquinone, 5
binding, 2
biodegradability, 2, 3, 61
biodegradable, 10
biodegradation, 2, 55, 60
biological, vii, 2, 4, 23, 67
biological processes, 3
biology, 24
biomass, 3
black, 2, 35
bleaching, 11, 17, 36, 40, 42, 45, 49, 50, 53, 55
bonds, 39, 45
British, 59
bubble, 28
bubbles, 28
by-products, 2, 12, 13, 23, 39, 41, 42, 43, 45, 46, 50

C

capacity, 3
capital, 53
capital cost, 53
carbon, 3, 6, 17, 20, 34, 61
carboxylates, 37
carcinogenic, vii, 1, 4
carcinogenicity, 2
catalyst, 6, 10, 18, 20, 34, 63, 64, 66
catalysts, 10, 12, 13, 31, 63, 71
catalytic, 6, 9, 10, 11, 12, 37
catechol, 5
cation, 29
cations, 31
cavitation, 28
cavities, 31
C-C, 64
cellulose, 3
channels, 26, 46
charged particle, 29
chemical, vii, 2, 4, 5, 6, 21, 22, 23, 25, 27, 29, 31, 55, 64, 67, 68
chemical oxidation, 64
chemical reactions, 31, 68
chemicals, 2, 23, 32

chemistry, 24, 42, 61, 67
chlorinated hydrocarbons, 3
chlorine, 3, 21
chlorobenzene, 5
chloroform, 4
chlorophenol, 5
chromatography, 71
classical, 4, 5, 22, 26, 45
classification, 7
classified, vii, 2, 6, 13, 22, 29
clays, 10
cleavage, 2, 13, 20, 22, 39, 42, 45
CO_2, 4, 46
coagulation, 3, 61
cobalt, 10, 64
coconut, 3
commercial, 2, 19, 32, 55, 60, 62, 63, 65
community, 4
components, 3, 17
composites, 63
compounds, vii, 1, 9, 15, 21, 28, 36, 62, 63, 67, 68, 72
concentration, vii, 1, 11, 15, 16, 17, 19, 22, 27, 32, 33, 34, 37, 38, 39, 40, 41, 42, 46, 48, 49, 53, 64
conditioning, 31, 54
conductance, 18, 31
conduction, 5, 19
conductive, 29, 70
conductivity, 32, 34
configuration, 21, 26, 42, 53
congress, iv, 68
consumption, 4, 36, 37
contaminant, 15
contaminants, 2, 3, 4, 55
control, 16
cooling, 28
corona, 23, 24, 25, 26, 45, 68, 69
corona discharge, 23, 24, 25, 26, 45, 68, 69
cost-effective, viii, 32
costs, 15, 53, 54, 55
cotton, 1, 59
Croatia, 32, 57
crystalline, 31
crystalline solids, 31

Index

cyanosis, 2

D

decay, 15, 29
decomposition, 17, 21, 22, 28, 65, 67
degradation, 2, 3, 9, 10, 11, 13, 16, 17, 19, 21, 22, 25, 26, 28, 29, 31, 34, 35, 36, 37, 38, 39, 40, 42, 43, 44, 45, 46, 47, 49, 50, 51, 52, 54, 55, 60, 61, 62, 63, 64, 65, 66, 67, 68, 69, 70, 71
degradation mechanism, 22, 39, 42, 45
degree, 15, 41
demand, 5
dermatitis, 59
destruction, 42
detergents, 71
detoxification, 11, 16
detoxifying, 62
diagnostic, 28
diagnostic ultrasound, 28
dielectric, 24
diffraction, 13
disability, vii
disabled, 36
discharges, 26, 68, 69
disinfection, 14
dispersion, 19
dissociation, 22
dissolved oxygen, 19
dizziness, 2
doped, 10, 12
dosage, 19, 35
drinking, 16, 20, 24
drinking water, 16, 20
duration, 33
dyeing, 1, 70
dyes, vii, 1, 2, 3, 11, 13, 15, 17, 19, 22, 26, 28, 29, 45, 55, 59, 60, 61, 62, 63, 64, 65, 66, 68, 69, 70

E

earth, 20

ecological, 5, 12, 17, 39, 41
ecosystems, 2
Education, 57
effluent, vii, 4, 23, 53, 67, 68, 71
effluents, 1, 2, 55, 59, 64, 68
Einstein, 14, 16, 33
electric field, 23
electric potential, 24
electrical, vii, 6, 20, 23, 25, 62, 68, 69, 71
electricity, 53
electrochemical, 27, 28, 29, 70
electrochemistry, 24
electrodes, 24, 26, 29, 66, 70
electrolyte, 70
electrolytes, 29
electromagnetic, 14, 29
electron, 3, 18, 19, 21, 29, 62, 63
electron beam, 29
electronic, iv, 14, 21
electrons, 18, 25, 29, 68
electrostatic, iv
energy, 6, 13, 14, 16, 18, 25, 29, 70
engineering, 67
environment, vii, 2, 28
environmental, 1, 61, 62, 66, 69, 70
enzymatic, 68
equipment, 4
ethanol, 5
ethylene, 5
ethylene glycol, 5
Europe, 20
European, 60
everyday life, 1
evidence, 63
excitation, 18, 29
expert, iv
exposure, 3

F

ferric ion, 12
ferrous ion, 10, 12, 64
film, 35, 66
filtration, 3, 16, 55
financial support, 57

flexibility, 5
flocculation, 3
flow, 33, 53
flow rate, 33, 53
fluid, 28
fluorescence, 14
food, 1
formaldehyde, 5
fractionation, 65
free radical, 27
free radicals, 27

G

gas, 21, 22, 23, 26, 28, 33, 45, 62, 69, 71
gas chromatograph, 71
gas phase, 21, 23, 26, 45
gases, 16
generation, vii, 3, 4, 6, 9, 15, 17, 18, 21, 22, 25, 26, 27, 31, 33, 46
generators, 20
genotoxic, 2
Georgia, 33
Germany, 20, 34
glass, 33
glycerol, 5
glycol, 5
ground water, 16

H

H_2, 25, 33
handling, vii, 1
hazards, 2
health, vii, 1, 2
heat, 28
heating, 28
heavy metal, 15
heavy metals, 15
heme, 2
hemoglobin, 2
heterogeneous, 10, 12, 63, 64, 65, 66
heterogeneous systems, 66
homogeneous, 9, 64, 65, 66
homogenous, 10, 12
hospital, 20
hospitals, 20
host, 31
human, vii, 1, 2
hybrid, 21, 26, 27, 45, 61, 69, 71
hydro, 3, 20
hydrocarbon, 21
hydrocarbons, 3, 20
hydrogen, 6, 9, 11, 12, 27, 29, 32, 33, 49, 61, 63, 64, 65, 66, 67, 68, 71
hydrogen atoms, 29
hydrogen peroxide, 6, 9, 11, 12, 29, 32, 33, 49, 61, 64, 65, 66, 67, 68, 71
hydrolysis, 1, 59
hydroquinone, 5
hydrothermal, 6
hydroxide, 32
hydroxyl, vii, 4, 5, 18, 21, 27, 29, 49, 68

I

immobilization, 60
impurities, 2
in situ, 22, 67
inactive, 3
industrial, 3, 61, 62, 66, 69, 70
industrial application, 61, 62, 66, 69, 70
industry, 59
inert, 35
influence, vii, 6, 20, 25, 27, 31, 32, 33, 44, 45, 46, 47, 49, 50, 51, 52
infrared, 13
inhibition, 37, 38, 41
inhibitor, 21
initiation, 13
injury, iv
inorganic, 15, 20, 21, 28, 72
input, 28
intensity, 42
interest, 4
interface, 63
interference, 13, 36, 40
ion exchangers, 31
ionic, 26

ionization, 29
ionizing radiation, 29
ions, 6, 9, 10, 12, 18, 21, 25, 27, 29, 31, 35, 54
iron, 2, 6, 10, 12, 16, 31, 32, 33, 34, 35, 54, 62, 63, 64, 69, 71
irradiation, 6, 11, 12, 13, 14, 23, 27, 29, 33, 36, 39, 45, 46, 53, 54, 61, 63, 64, 70

J

Japan, 34

K

kinetic model, 65, 66, 67
kinetics, 4, 22, 50, 61, 63, 67

L

laws, 13
leaching, 35
lead, 46
lifetime, 14, 25
ligands, 10
light beam, 13
likelihood, 2
limitation, vii, 15, 16
limitations, 4, 11, 16
linear, 45, 53
liquid phase, 3, 22, 23, 25, 26, 45
literature, 3, 11, 14, 16, 17, 34, 36, 42, 55
London, 61, 64, 65, 66, 69, 70

M

magnetic, iv, 32, 33
maintenance, 53, 54
malic, 5
manganese, 4, 10
manufacturing, vii, 1
market, 53
mass, 22, 71
mass spectrometry, 71
mass transfer, 22
measurement, 71
mechanical, iv, vii, 2, 4, 6
media, 64, 65, 70
mercury, 14
metabolic, 60
metal ions, 10, 32
metals, 32
methanol, 5, 21
methemoglobinemia, 2
methodology, 2
Microbial, 60
microorganism, 3
microorganisms, 23, 55
migration, 25
mineralization, vii, 4, 10, 11, 13, 15, 17, 20, 23, 30, 31, 32, 33, 34, 37, 38, 41, 43, 44, 45, 46, 47, 48, 49, 50, 51, 52, 53, 54, 55, 61, 65, 70
mineralized, 15, 34, 42, 53, 55
minerals, 3, 60
mixing, 33
mobility, 25
molar ratio, 32
molar ratios, 32
molecular structure, 19, 20, 23, 35, 36
molecules, 14, 15, 20, 25, 28, 29, 31, 36, 38, 42, 45, 46, 55
monitoring, 35, 36, 37, 39, 40, 41
mutagenic, 1

N

Na2SO4, 29
NaCl, 29
nanocrystalline, 19
National Science Foundation, 57
natural, 1, 3
New York, iii, iv, 59, 60, 61, 62, 65, 66, 69, 70
Newton, 13
Ni, 63
nonlinear, 28
non-thermal, 6, 25
non-uniform, 24

nuclei, 29

O

observations, 13
oils, 15
optimization, 33, 49, 62
organic, vii, 1, 3, 4, 9, 10, 11, 13, 15, 17, 21, 22, 23, 28, 29, 32, 34, 36, 38, 41, 43, 45, 46, 50, 53, 55, 62, 63, 66, 67, 68, 69, 70, 72
organic compounds, vii, 1, 9, 16, 21, 67, 68, 72
organic matter, 10, 17
oxalic, 4, 13, 37, 46
oxalic acid, 4
oxidants, 13, 18
oxidation, vii, 4, 6, 7, 9, 11, 12, 13, 14, 18, 20, 21, 29, 34, 46, 61, 62, 63, 64, 65, 66, 67, 68, 69, 70, 71
oxidative, 4, 17, 35, 45
oxide, 4, 64
oxides, 10
oxygen, 2, 5, 19, 20, 23, 31, 33
ozonation, 20, 22, 61, 68
ozone, 3, 5, 6, 9, 16, 17, 20, 21, 22, 23, 26, 31, 33, 42, 45, 46, 53, 54, 57, 61, 65, 66, 67, 68, 69, 71, 72

P

paper, 1, 17, 19, 27
parameter, 5, 17, 19, 39, 41, 49
particles, vii, 16, 19, 29, 32, 33, 38, 67
penicillin, 67
periodic, 28
peroxide, 5, 6, 9, 11, 12, 29, 32, 33, 49, 61, 63, 64, 65, 66, 67, 68, 69, 71
pH, vii, 11, 16, 17, 19, 21, 23, 26, 30, 32, 33, 34, 35, 42, 43, 44, 45, 47, 48, 49, 50, 51, 52, 54, 64
pH values, vii, 23, 32, 33, 34, 42, 43, 44, 47, 48, 49, 51, 52, 54
pharmaceutical, 1
phenol, 5, 39, 42, 62, 63, 64, 65, 69, 71

phosphorescence, 14
photocatalysis, 18, 19, 66
photocatalysts, 19
Photocatalytic, 18, 66, 67, 71
photochemical, vii, 6, 13, 14, 16, 42, 55, 65, 66, 68
photodegradation, 64
photolysis, 12, 13, 14, 15, 16, 17, 20, 37, 46, 49, 64
photon, 33
photonic, 18, 71
photons, 14, 18
photooxidation, 71
photosensitivity, 15
physical mechanisms, 28
physical properties, 24
piezoelectric, 28
pigments, 1
pinhole, 69
Planck constant, 14
plants, 20
plasma, 6, 25, 62
pollutant, vii, 10, 11, 29, 31, 32, 34, 38, 55, 62, 63, 66
pollutants, vii, 2, 4, 10, 13, 15, 17, 22, 53, 55, 61, 62, 66, 69, 70
pollution, 1, 2
poor, 2, 3, 15
pore, 3, 31, 46
pores, 23, 38, 46
porous, 22
potassium, 3, 16
powder, 10, 12, 19, 31, 32, 33, 34, 35, 54, 63, 71
power, vii, 4, 11, 28, 34, 70
powers, 28
precipitation, 4
prediction, 61
preparation, iv
pressure, 6, 14, 28
priorities, 59
production, vii, 1, 2, 3, 20, 24, 25, 29, 50, 61
promote, 21
promoter, 21
property, iv

protons, 29
pseudo, 22
pulse, 69, 70
pulses, 25
purification, 15, 20, 62

Q

quantum, 16
quinone, 62

R

radiation, 6, 14, 15, 16, 17, 18, 20, 29, 42, 61, 66, 70, 71
radical, 5, 6, 10, 13, 15, 17, 19, 22, 23, 25, 26, 29, 31, 34, 38, 42, 43, 46, 49, 65
radical mechanism, 10, 42
range, vii, 3, 6, 10, 14, 19, 21, 34, 42, 49
reaction mechanism, 14
reaction rate, 46, 65
reaction rate constants, 46, 65
reaction time, 11
reactivity, 4, 9, 12, 15, 21, 22, 63
reagent, 9, 11, 21, 53, 62
reagents, 11, 53
recombination, 20
recovery, 62
redox, 5, 21
reduction, 12, 16, 19, 21, 29, 62, 70, 71
regeneration, 3
regular, 31
regulations, vii, 2
relationships, 2
research, vii, 4, 31, 32, 34, 55
resin, 63
resins, 10, 12
returns, 14
rhinitis, 59
rice, 3
rice husk, 3
rings, 36, 38, 39, 42
room temperature, 5, 14, 32, 34

S

safeguard, vii, 2
salts, 3, 10, 12, 16
scavenger, 16, 21, 35, 49
scientific, 4, 55
scientific community, 4
sedimentation, 3
selectivity, vii, 4
semiconductor, 18, 67
semiconductors, 19
series, 26, 45
services, iv
shuttles, 62
silica, 31
sludge, 3
sodium, 32
sodium hydroxide, 32
solar, 6, 54, 64
solubility, 1
solutions, vii, 15, 22, 67
sorbents, 60
species, vii, 4, 5, 21, 23, 25, 26, 29, 42, 63, 68, 69
spectra, 13, 34, 35, 37, 38, 40
spectrum, 13
spheres, 1
substances, vii, 1, 56
sugarcane, 3
sulfuric acid, 32
Sun, 68, 69
sunlight, 1
supercritical, 6
supply, 31
surface area, 46
surfactants, 71
susceptibility, 13
suspensions, 64, 67
symptoms, 2
synthesis, 61
synthetic, vii, 1, 3, 27, 31, 32, 33, 38, 45, 46, 50, 57, 60, 61, 64, 69
systems, 2, 11, 17, 66

T

technology, 64
temperature, 6, 11, 14, 22, 28, 32, 34
tetrachloroethane, 4
tetrahydrofuran, 5
textile, 1, 3, 59, 60, 65, 66, 67, 70
theoretical, 69
thermal, 69
thin film, 19
threat, 2
three-dimensional, 31
threshold, 11
time, 11, 17, 34, 36, 45, 46, 50, 54
TiO_2, 6, 13, 18, 19, 64, 66, 67, 68, 71
titanium, 67
titanium dioxide, 67
toluene, 5
total organic carbon, 6, 34
total organic carbon (TOC), viii, 6, 11, 15, 23, 32, 33, 34, 35, 37, 38, 39, 41, 42, 43, 45, 46, 49, 50, 53, 54
toxic, vii, 1, 4, 16, 28, 55
toxic products, 28
toxicity, 2, 59, 70
transducer, 28
transfer, 3, 18, 21, 22, 26, 29, 45, 63, 65, 67
transformations, 3
transmission, 15, 24
transparency, 45
treatment methods, vii, 2, 4, 5, 13, 22, 55
trend, 50, 53

U

ultrasonic waves, 28
ultrasound, 6, 20, 23, 27, 28, 70
ultraviolet irradiation, 61
uniform, 24
users, vii, 2
ultraviolet (UV), vii, 6, 9, 11, 12, 13, 14, 15, 16, 17, 19, 20, 22, 23, 31, 32, 33, 34, 35, 37, 38, 39, 40, 41, 42, 43, 44, 45, 47, 48, 49, 50, 51, 52, 53, 54, 55, 61, 63, 64, 65, 66, 67, 68, 71
UV irradiation, 11, 12, 13, 23, 33, 39, 45, 46, 53, 54, 63
UV light, 12, 13, 31, 36, 38
UV radiation, 14, 15, 16, 17, 20, 42, 61, 66

V

vacuum, 14
valence, 18
values, vii, 23, 32, 33, 34, 42, 43, 44, 47, 48, 49, 51, 52, 53, 54, 55
vapor, 14, 28
variable, 33
visible, 1, 13, 67

W

Washington, 71
waste, vii, 3, 4, 55, 59
waste water, 59
wastes, 61, 62, 66, 69, 70
wastewater, vii, 2, 4, 5, 9, 10, 11, 13, 15, 17, 19, 23, 26, 27, 28, 29, 31, 32, 34, 37, 40, 42, 45, 49, 50, 53, 55, 60, 61, 64, 65, 66, 69, 70, 71
wastewater treatment, vii, 2, 4, 5, 27, 61, 64, 65, 66, 69, 70, 71
wastewaters, vii, 1, 3, 5, 6, 12, 13, 16, 17, 22, 24, 27, 45, 55, 60, 62, 67, 70
water, 1, 3, 4, 6, 9, 10, 13, 15, 16, 17, 19, 20, 21, 22, 23, 25, 26, 27, 28, 29, 31, 32, 33, 46, 55, 61, 62, 64, 65, 66, 67, 68, 69, 70, 71, 72
weakness, 2
wet, 6, 12, 63
wool, 1
workers, 59

X

xylene, 5

Y

yield, 11, 14, 16

Z

zeolites, vii, 3, 10, 12, 27, 31, 32, 33, 38, 39, 44, 45, 46, 47, 50, 51, 52, 57, 69, 70, 71

ZnO, 6, 13, 19, 66, 67